ECONOMETRICS

STATISTICS

Textbooks and Monographs

A SERIES EDITED BY

D. B. OWEN, Coordinating Editor
Department of Statistics
Southern Methodist University
Dallas, Texas

PAUL D. MINTON
Virginia Commonwealth University
Richmond, Virginia

JOHN W. PRATT
Harvard University
Boston, Massachusetts

OTHER VOLUMES IN PREPARATION

ECONOMETRICS

PETER SCHMIDT

Department of Economics
University of North Carolina at Chapel Hill
Chapel Hill, North Carolina

CRC Press
Taylor & Francis Group
Boca Raton London New York

CRC Press is an imprint of the
Taylor & Francis Group, an **informa** business

CRC Press
Taylor & Francis Group
6000 Broken Sound Parkway NW, Suite 300
Boca Raton, FL 33487-2742

First issued in paperback 2019

ISBN-13: 978-0-8247-8735-6 (hbk)
ISBN-13: 978-0-367-40344-7 (pbk)

Visit the Taylor & Francis Web site at
http://www.taylorandfrancis.com

and the CRC Press Web site at
http://www.crcpress.com

CV 09.05.2019 1547

PREFACE

This book is essentially a collection of proofs of fundamental econometric results. For some time it has appeared to me that there has been a need for such a book, since, although there are now several good econometrics texts available, they all tend to omit proofs of the more difficult or obscure theorems. The reader of these texts is referred to a variety of scholarly journals and monographs (some of them out of print) for these proofs, and he is likely to be bothered both by the difficulty of locating the sources and by differences in notation. The idea behind this book is to collect these proofs in one place and in one notation.

The distinguishing feature of the book, therefore, is that there are no omitted proofs; every result that is stated is proved. (The only exception to this rule is the Appendix, which states without proof certain results from the statistical literature.) As a result, the topics covered in the book are roughly those covered in most of the texts used in the typical graduate-level course. There is a somewhat deeper coverage of some topics, notably simultaneous equations models, but the main difference between this book and such texts is the level of rigor.

While the book is logically self-contained, at least as far as its econometric content is concerned, it is not recommended as a first text in econometrics. In order to keep the book relatively short, much of the discussion and motivation that a text usually contains has been omitted. It is assumed that the reader is familiar with most of the results, and with their place in the econometric literature, and is interested in this book primarily for their proofs. The other prerequisite to understanding the material is a reasonably good knowledge of calculus, matrix algebra, and classical statistics.

The matrix algebra in particular is sometimes rather advanced; this is unavoidable in a book of this level.

It is hoped that this book will prove useful to graduate students and to practicing econometricians. Specifically, it may be useful as a supplementary text in beginning graduate-level econometrics courses; as a text in more advanced courses and seminars; and as a reference.

A large number of colleagues, students and friends assisted in one way or another in the preparation of this book. I am especially grateful to my former teacher, Jan Kmenta, for his continuing interest in my understanding of econometrics. I would also like to thank June Maxwell for her exceptional care in preparing the final typed copy. Finally, I would like to thank my wife for her help, encouragement and patience throughout the years in which I worked on this book.

Contents

ECONOMETRICS

Chapter 1

LEAST SQUARES UNDER IDEAL CONDITIONS

1.1 INTRODUCTION

Consider the linear regression model

$$y_t = \beta_1 X_{t1} + \beta_2 X_{t2} + \cdots + \beta_K X_{tK} + \varepsilon_t \qquad (t = 1, 2, \cdots, T)$$

where y_t is the t-th observation on the dependent variable in the regression, X_{ti} is the t-th observation on the i-th independent variable (regressor), β_i is the regression coefficient corresponding to the i-th regressor, and ε_t is the t-th observation on the disturbance (error) term. Note that T is the number of observations and K is the number of regressors.

The above model can of course be written in matrix form as

$$y = X\beta + \varepsilon$$

Here y is of dimension $T \times 1$, X is $T \times K$, β is $K \times 1$, and ε is $T \times 1$.

1. DEFINITION: The model $y = X\beta + \varepsilon$ is said to satisfy the *full ideal conditions* [see Note 1 at end of chapter] if:

(i) X is a nonstochastic matrix of rank $K < T$, and has the property that

$$\lim_{T\to\infty} \frac{X'X}{T} \equiv Q$$

exists as a finite nonsingular matrix

(ii) ε has a multivariate normal distribution with mean vector zero and covariance matrix $\sigma^2 I$.

2. COMMENT: The following condition will also sometimes be imposed:

(iii) One of the regressors is a constant term; that is, $X_{ti} = 1$ for all t and some $1 \le i \le K$.

It will be specifically noted whenever a result requires this condition in addition to the full ideal conditions (as given in Definition 1).

1.2 LEAST SQUARES: ALGEBRAIC RESULTS

This section will deal with results of least squares estimation which are purely algebraic in character. They do not depend on any assumptions about the vector ε of disturbances.

1. THEOREM: The ordinary least squares (OLS) estimator of β in the model $y = X\beta + \varepsilon$ is

$$\hat{\beta} = (X'X)^{-1}X'y$$

Proof: By definition, the OLS estimators are those values of $\beta_1, \cdots, \beta_K$ which minimize the sum of squared errors

$$\sum_{t=1}^{T} (y_t - \beta_1 X_{t1} - \cdots - \beta_K X_{tK})^2$$

In matrix notation, the sum of squared errors is

$$(y - X\beta)'(y - X\beta) = y'y - 2\beta'X'y + \beta'X'X\beta$$

Differentiating with respect to β and setting the derivative equal to zero yields

$$0 = -2X'y + 2X'X\hat{\beta} \quad \text{[see Note 1]}$$

or

$$X'X\hat{\beta} = X'y$$

a set of equations referred to as the least squares normal equations.

Since X is of rank K, $(X'X)^{-1}$ exists and hence we have

$$\hat{\beta} = (X'X)^{-1}X'y$$

as claimed. That this does indeed correspond to a minimum of the sum of squared errors can be verified by noting that the matrix of second-order partial derivatives with respect to β is just $2X'X$, a positive definite matrix. Alternatively, note that

$$(y - X\beta)'(y - X\beta) = (y - X\hat{\beta})'(y - X\hat{\beta}) + (\hat{\beta} - \beta)'X'X(\hat{\beta} - \beta)$$

which is clearly minimized by taking $\beta = \hat{\beta}$.

2. DEFINITION: The vector of least squares *residuals* is defined by

$$e = y - X\hat{\beta}$$

3. PROPOSITION: $X'e = 0$.

Proof: $X'e = X'(y - X\hat{\beta}) = X'y - X'X(X'X)^{-1}X'y = 0.$

4. COROLLARY: If one of the regressors is a constant term, then the sum of the residuals is zero.

5. NOTATION:

$$\hat{y} = x\hat{\beta}$$

$$SSE = e'e = \sum_{t=1}^{T} (y_t - \hat{y}_t)^2$$

$$SSR = \sum_{t=1}^{T} (\hat{y}_t - \bar{y})^2$$

$$SST = \sum_{t=1}^{T} (y_t - \bar{y})^2$$

where

$$\bar{y} = \frac{1}{T} \sum_{t=1}^{T} y_t$$

6. LEMMA: $e'e = y'y - \hat{y}'\hat{y}$.

Proof: $e'e = y'y - 2\hat{\beta}'X'y + \hat{\beta}'X'X\hat{\beta}$. But since $X'y = X'X\hat{\beta}$, $e'e = y'y - 2\hat{\beta}'X'X\hat{\beta} + \hat{\beta}'X'X\hat{\beta} = y'y - \hat{y}'\hat{y}$.

7. PROPOSITION: If one of the regressors is a constant term,

$$SST = SSR + SSE$$

Proof: From Lemma 6, $y'y = \hat{y}'\hat{y} + e'e$. Therefore

$$y'y - T\bar{y}^2 = \hat{y}'\hat{y} - T\bar{y}^2 + e'e \quad \text{or}$$

$$SST = \hat{y}'\hat{y} - T\bar{y}^2 + SSE$$

Hence what must be shown is that $SSR = \hat{y}'\hat{y} - T\bar{y}^2$. But

$$SSR = \sum_{t=1}^{T} (\hat{y}_t - \bar{y})^2 = \hat{y}'\hat{y} - 2\sum_{t=1}^{T} \hat{y}_t\bar{y} + T\bar{y}^2$$

Furthermore, if one of the regressors is a constant term,

$$\sum_{t=1}^{T} \hat{y}_t\bar{y} = \sum_{t=1}^{T} y_t\bar{y} - \sum_{t=1}^{T} e_t\bar{y} = \sum_{t=1}^{T} y_t\bar{y} = T\bar{y}^2$$

Therefore $SSR = \hat{y}'\hat{y} - 2T\bar{y}^2 + T\bar{y}^2 = \hat{y}'\hat{y} - T\bar{y}^2$ and $SST = SSR + SSE$.

8. DEFINITION: $R^2 = 1 - \frac{SSE}{SST}$.

9. PROPOSITION: If one of the regressors is a constant term, then $0 \le R^2 \le 1$.

Proof: The result follows from the fact that $SST = SSR + SSE$, both SSR and SSE being nonnegative by definition.

10. PROPOSITION: If one of the regressors is a constant term, then

$$R^2 = \frac{SSR}{SST}$$

Proof: $\frac{SSR}{SST} = \frac{SST - SSE}{SST} = 1 - \frac{SSE}{SST} = R^2$.

It is worth noting that unless one of the regressors is a constant term, Propositions 7, 9, and 10 need not hold. In particular, R^2 will always be less than or equal to one, but it need not be nonnegative unless one of the regressors is a constant term [see Note 2].

11. PROPOSITION: If one of the regressors is a constant term, then the (empirical) correlation coefficient between y and $\hat{y}$ is $\sqrt{R^2}$.

Proof: The empirical variance of y is

$$V(y) = \frac{1}{T}\sum_{t=1}^{T}(y_t - \bar{y})^2 = \frac{1}{T}\,SST$$

The empirical variance of $\hat{y}$ is

$$V(\hat{y}) = \frac{1}{T}\sum_{t=1}^{T}(\hat{y}_t - \bar{y})^2 = \frac{1}{T}\,SSR$$

since $\bar{y} = \frac{1}{T}\sum_{t=1}^{T}\hat{y}_t$ if one of the regressors is a constant term.

Finally, the empirical covariance between y and $\hat{y}$ is

$$\mathrm{cov}(y,\hat{y}) = \frac{1}{T}\sum_{t=1}^{T}(y_t - \bar{y})(\hat{y}_t - \bar{y})$$

$$= \frac{1}{T}\sum_{t=1}^{T}(y_t\hat{y}_t - \bar{y}\hat{y}_t - \bar{y}y_t + \bar{y}^2)$$

But $\sum_{t=1}^{T}y_t\hat{y}_t = \sum_{t=1}^{T}\hat{y}_t^2$ since $\hat{y}'e = \hat{\beta}'X'e = 0$, and $\sum_{t=1}^{T}\bar{y}\hat{y}_t = \sum_{t=1}^{T}\bar{y}y_t = T\bar{y}^2$. Therefore

$$\mathrm{cov}(y,\hat{y}) = \frac{1}{T}\sum_{t=1}^{T}\hat{y}_t^2 - \bar{y}^2 = \frac{1}{T}\,SSR$$

The correlation coefficient between y and $\hat{y}$ is then

$$\frac{\mathrm{cov}(y,\hat{y})}{\sqrt{V(y)V(\hat{y})}} = \frac{\sqrt{SSR}}{\sqrt{SST}} = \sqrt{R^2}$$

Because of this result, R^2 is sometimes referred to as the (squared) multiple correlation coefficient between y and X.

1.3 BASIC STATISTICAL RESULTS

This section will deal with certain basic statistical results of least squares estimation. The full ideal conditions will be assumed, except that the assumption of normality of the disturbances will not be used. As before, $\hat{\beta}$ denotes the OLS estimator of β.

Estimation of β

1. LEMMA: $\hat{\beta} = \beta + (X'X)^{-1}X'\varepsilon$.

Proof: $\hat{\beta} = (X'X)^{-1}X'y = (X'X)^{-1}X'(X\beta + \varepsilon) = \beta + (X'X)^{-1}X'\varepsilon$.

2. PROPOSITION: $\hat{\beta}$ is an unbiased estimator of β.

Proof:

$$
\begin{aligned}
E(\hat{\beta}) &= \beta + E[(X'X)^{-1}X'\varepsilon] \\
&= \beta + (X'X)^{-1}X'E(\varepsilon) && \text{[since X is nonstochastic]} \\
&= \beta && \text{[since } E(\varepsilon) = 0]
\end{aligned}
$$

3. PROPOSITION: The covariance matrix of $\hat{\beta}$ is $\sigma^2(X'X)^{-1}$.

Proof: By definition, the covariance matrix of $\hat{\beta}$ is

$$
\begin{aligned}
E(\hat{\beta} - \beta)(\hat{\beta} - \beta)' &= E[(X'X)^{-1}X'\varepsilon\varepsilon'X(X'X)^{-1}] \\
&= (X'X)^{-1}X'\sigma^2 IX(X'X)^{-1} \\
&= \sigma^2(X'X)^{-1}
\end{aligned}
$$

4. THEOREM (Gauss-Markov): The OLS estimator $\hat{\beta}$ is the best linear unbiased estimator (BLUE) of β.

Proof: Consider any estimator linear in y, say $\tilde{\beta} = Cy$. Let $C = (X'X)^{-1}X' + D$. Then

$$
\begin{aligned}
E(\tilde{\beta}) &= E[(X'X)^{-1}X' + D][X\beta + \varepsilon] \\
&= \beta + DX\beta
\end{aligned}
$$

so that for $\tilde{\beta}$ to be unbiased we require $DX = 0$.

Then the covariance matrix of $\tilde{\beta}$ is

$$
\begin{aligned}
E(\tilde{\beta} - \beta)(\tilde{\beta} - \beta)' &= E[(X'X)^{-1}X' + D]\varepsilon\varepsilon'[(X'X)^{-1}X' + D]' \\
&= \sigma^2[(X'X)^{-1}X'IX(X'X)^{-1} + DIX(X'X)^{-1} \\
&\quad + (X'X)^{-1}X'ID' + DID'] \\
&= \sigma^2(X'X)^{-1} + \sigma^2 DD' \qquad \text{since } DX = 0
\end{aligned}
$$

But DD' is a positive semidefinite matrix, which shows that the covariance matrix of $\tilde{\beta}$ equals the covariance matrix of $\hat{\beta}$ *plus* a positive semidefinite matrix. Hence $\hat{\beta}$ is efficient relative to any other linear unbiased estimator of β.

5. COROLLARY: Consider any linear combination $\lambda'\beta$ of the elements of β. Then the BLUE of $\lambda'\beta$ is $\lambda'\hat{\beta}$.

6. PROPOSITION: The OLS estimator $\hat{\beta}$ is consistent.

Proof: Denote $\lim_{T\to\infty}(X'X/T)$ by $\underset{\sim}{Q}$; by assumption, $\underset{\sim}{Q}$ is finite and nonsingular. Then $\lim_{T\to\infty}(X'X/T)^{-1} = \underset{\sim}{Q}^{-1}$ is also finite. Therefore

$$
\begin{aligned}
\lim_{T\to\infty}(X'X)^{-1} &= \lim_{T\to\infty} \frac{1}{T}\left(\frac{X'X}{T}\right)^{-1} \\
&= \lim_{T\to\infty} \frac{1}{T} \underset{\sim}{Q}^{-1} \\
&= 0
\end{aligned}
$$

Since $\hat{\beta}$ is unbiased and its covariance matrix vanishes asymptotically, it must be consistent.

An alternative proof of Proposition 6 makes use of the following useful lemma.

7. LEMMA: $\text{plim}(X'\varepsilon/T) = 0$.

Proof: $E\,\frac{X'\varepsilon}{T} = 0$. Also $E\left(\frac{X'\varepsilon}{T}\right)\left(\frac{X'\varepsilon}{T}\right)' = \frac{\sigma^2}{T}\frac{X'X}{T}$, so that

$$
\lim_{T\to\infty} E\left(\frac{X'\varepsilon}{T}\right)\left(\frac{X'\varepsilon}{T}\right)' = \lim_{T\to\infty} \frac{\sigma^2}{T} \underset{\sim}{Q} = 0
$$

But the facts that $E\,\frac{X'\varepsilon}{T} = 0$ and $\lim_{T\to\infty} E\left(\frac{X'\varepsilon}{T}\right)\left(\frac{X'\varepsilon}{T}\right)' = 0$ imply that $\text{plim}\,\frac{X'\varepsilon}{T} = 0$ [see Note 1].

Given Lemma 7, then, one can easily prove Proposition 6 by noting that

$$\hat{\beta} = \beta + (X'X)^{-1}X'\varepsilon = \beta + \left(\frac{X'X}{T}\right)^{-1}\frac{X'\varepsilon}{T}$$

therefore $\text{plim}\,\hat{\beta} = \beta + \underset{\sim}{Q}^{-1}\text{plim}(X'\varepsilon/T) = \beta$.

Estimation of σ^2

8. LEMMA: The matrix $M = I - X(X'X)^{-1}X'$ is symmetric ($M' = M$) and idempotent ($MM = M$). Furthermore, $MX = 0$.

9. LEMMA: $e = M\varepsilon$, where e is the residual vector, ε is the disturbance term vector, and M is defined in Lemma 8.

Proof: $e = y - X\hat{\beta} = y - X(X'X)^{-1}X'y = My = MX\beta + M\varepsilon = M\varepsilon$.

10. LEMMA: $SSE = \varepsilon'M\varepsilon$.

Proof:

$$\begin{aligned} SSE &= e'e = \varepsilon'M'M\varepsilon \\ &= \varepsilon'MM\varepsilon \quad \text{(since } M \text{ is symmetric)} \\ &= \varepsilon'M\varepsilon \quad \text{(since } M \text{ is idempotent)} \end{aligned}$$

11. THEOREM: $E(SSE) = \sigma^2(T - K)$.

Proof:

$$\begin{aligned} E(SSE) &= E(\varepsilon'M\varepsilon) \\ &= E(\text{trace } \varepsilon'M\varepsilon) \quad \text{(since } \varepsilon'M\varepsilon\text{, a scalar, equals its trace)} \\ &= E(\text{trace } M\varepsilon\varepsilon') \\ &= \text{trace } M\sigma^2 I \\ &= \sigma^2 \text{ trace } M \end{aligned}$$

But

$$\begin{aligned} \text{trace } M &= \text{trace } I_T - \text{trace } X(X'X)^{-1}X' \\ &= \text{trace } I_T - \text{trace}(X'X)^{-1}X'X \\ &= T - K \quad \text{[see Note 2].} \end{aligned}$$

12. COROLLARY: An unbiased estimator of σ^2 is

$$s^2 \equiv \frac{SSE}{T - K}$$

13. PROPOSITION: s^2 is a consistent estimator of σ^2.

Proof:

$$\begin{aligned}
\text{plim } s^2 &= \text{plim } \frac{\varepsilon' M \varepsilon}{T - K} = \text{plim } \frac{1}{T}\, \varepsilon' M \varepsilon \\
&= \text{plim } \frac{\varepsilon'\varepsilon}{T} - \text{plim}\left(\frac{\varepsilon' X}{T}\right)\left(\frac{X'X}{T}\right)^{-1}\left(\frac{X'\varepsilon}{T}\right) \\
&= \sigma^2 - 0 \cdot \underset{\sim}{Q}^{-1} \cdot 0 \qquad \text{(using Lemma 7)} \\
&= \sigma^2
\end{aligned}$$

Residuals and Forecast Errors

14. PROPOSITION: $E(e) = 0$.

Proof: $E(e) = E(y - X\hat{\beta}) = X\beta - X\beta = 0$.

15. PROPOSITION: $E(ee') = \sigma^2 M$.

Proof: $E(ee') = E(M\varepsilon\varepsilon' M') = M\sigma^2 IM = \sigma^2 M$ (using lemmas 8 and 9).

Now let us consider a set of T_o observations *not* included in the original sample of T observations. Specifically, let X_0 denote these T_o observations on the regressors, and y_0 these observations on y.

Now let y_0 be "forecasted" by

$$\hat{y}_0 = X_0\hat{\beta}$$

where $\hat{\beta} = (X'X)^{-1}X'y$ is the OLS estimator based on the original T observations. Finally, let v_o be the set of *forecast errors* defined by

$$v_o = y_0 - X_0\hat{\beta} = y_0 - X_0(X'X)^{-1}X'y$$

16. PROPOSITION: $E(v_o) = 0$.

Proof: $E(v_o) = E(y_o - X_o\hat{\beta}) = X_o\beta - X_o\beta = 0$.

17. PROPOSITION: $E(v_o v_o') = \sigma^2[I + X_0(X'X)^{-1}X_0']$.

Proof:

$$\begin{aligned} E(v_o v_o') &= E(y_0 - X_0\hat\beta)(y_0 - X_0\hat\beta)' \\ &= E(X_0\beta + \varepsilon_o - X_0\hat\beta)(X_0\beta + \varepsilon_o - X_0\hat\beta)' \\ &= X_0 E(\hat\beta - \beta)(\hat\beta - \beta)'X_0' + E(\varepsilon_o\varepsilon_o') \end{aligned}$$

(since ε_o and $\hat\beta - \beta$ are independent)

$$= \sigma^2 X_0(X'X)^{-1}X_0' + \sigma^2 I$$

It is perhaps worth noting that $\hat y_0 = X_0\hat\beta$ can be viewed both as an *estimator* of $E(y_0) = X_0\beta$ and as a *predictor* of y_0. Proposition 17 has established that the mean square error of prediction of y_0 is $\sigma^2[I + X_0(X'X)^{-1}X_0']$, which is the sum of the variance of ε_o plus the variance of $X_0\hat\beta$ as an estimator of $X_0\beta$. Since $X_0\hat\beta$ is the best linear unbiased estimator of $X_0\beta$ (according to Corollary 5), $\hat y_0 = X_0\hat\beta$ is sometimes referred to as the "best linear unbiased predictor" of y_0.

Finally, the following result shows that the forecast errors which one obtains outside the original sample are just a linear combination of the residuals which one would have obtained had the regression been carried out on the combined sample. It is actually an algebraic result, not a statistical one.

18. PROPOSITION: Let $y_* = \begin{bmatrix} y \\ y_0 \end{bmatrix}$, $X_* = \begin{bmatrix} X \\ X_0 \end{bmatrix}$, $\hat\beta_* = (X_*'X_*)^{-1}X_*'y_*$, and $e_o = y_0 - X_0\hat\beta_*$; as above, let $\hat\beta = (X'X)^{-1}X'y$ and $v_o = y_0 - X_0\hat\beta$. Then

$$v_o = [I - X_0(X_*'X_*)^{-1}X_0']^{-1}e_o$$

Proof:

$$\begin{aligned} e_o &= y_0 - X_0\hat\beta_* \\ &= y_0 - X_0(X_*'X_*)^{-1}X_*'y_* \\ &= y_0 - X_0(X_*'X_*)^{-1}(X'y + X_0'y_o) \end{aligned}$$

But $X'y = X'X\hat\beta$ and $X_0'y_0 = X_0'X_0\hat\beta + X_0'v_o$, and so we have

$$\begin{aligned} e_o &= y_0 - X_0(X_*'X_*)^{-1}(X'X\hat{\beta} + X_0'X_0\hat{\beta} + X_0'v_0) \\ &= y_0 - X_0(X_*'X_*)^{-1}(X'X + X_0'X_0)\hat{\beta} - X_0(X_*'X_*)^{-1}X_0'v_o \\ &= y_0 - X_0\hat{\beta} - X_0(X_*'X_*)^{-1}X_0'v_o \\ &= [I - X_0(X_*'X_*)^{-1}X_0']v_o \end{aligned}$$

1.4 FURTHER STATISTICAL RESULTS

This section will deal with further results of least squares estimation theory. It will assume all of the full ideal conditions, including the assumption of normality of the disturbances.

Distribution of $\hat{\beta}$ *and* s^2

1. THEOREM: $\hat{\beta}$ has a multivariate normal distribution with mean β and covariance matrix $\sigma^2(X'X)^{-1}$.

Proof: Propositions 1 and 2 of Section 1.3 established the mean and covariance matrix of $\hat{\beta}$. The normality of $\hat{\beta}$ is clear since $\hat{\beta}$ is linear in the multivariate normal vector y.

2. LEMMA: If Q is a symmetric, idempotent $T \times T$ matrix and ε is distributed as $N(0,\sigma^2 I_T)$, then

$$\frac{\varepsilon'Q\varepsilon}{\sigma^2}$$

(a scalar) is distributed as chi-squared with trace Q degrees of freedom.

Proof: By the symmetry of Q there exists an orthogonal matrix P such that $P'QP = L$, L being the diagonal matrix of eigenvalues of Q. Since Q is idempotent, trace Q of its eigenvalues equal one and the rest equal zero. Therefore

$$P'QP = L = \begin{bmatrix} I_* & 0 \\ 0 & 0 \end{bmatrix}$$

the dimension of the identity matrix I_* being trace Q.

Now define $v = P'\varepsilon$. Then $E(v) = P'E(\varepsilon) = 0$, and $E(vv') = P'E(\varepsilon\varepsilon')P = \sigma^2 I_T$ since P is orthogonal. Therefore v is distributed as $N(0,\sigma^2 I_T)$.

But now

$$\frac{\varepsilon'Q\varepsilon}{\sigma^2} = \frac{v'P'QPv}{\sigma^2} \qquad [\text{since } \varepsilon = (P')^{-1}v = Pv]$$

$$= \frac{1}{\sigma^2} v' \begin{bmatrix} I_* & 0 \\ 0 & 0 \end{bmatrix} v$$

$$= \frac{1}{\sigma^2} \sum_{i=1}^{\text{tr}Q} v_i^2$$

$$= \sum_{i=1}^{\text{tr}Q} \left(\frac{v_i}{\sigma}\right)^2$$

and this has a chi-squared distribution with trace Q degrees of freedom because it is the sum of squares of trace Q independent N(0,1) variables.

3. THEOREM: $(T - K)s^2/\sigma^2$ has a chi-squared distribution with T - K degrees of freedom.

Proof: $\frac{(T - K)s^2}{\sigma^2} = \frac{SSE}{\sigma^2} = \frac{\varepsilon'M\varepsilon}{\sigma^2}$, where $M = I - X(X'X)^{-1}X'$ is symmetric and idempotent. Lemma 2 therefore implies that $(T - K)s^2/\sigma^2$ has a chi-squared distribution with trace M degrees of freedom. But trace M = T - K [see Note 1], and so the result follows.

4. COROLLARY: The variance of s^2 is $2\sigma^4/(T - K)$.

Proof: The result follows immediately from the fact that $(T - K)s^2/\sigma^2$, a chi-squared variable with T - K degrees of freedom, has variance 2(T - K).

5. LEMMA: Let Q be a symmetric, idempotent T × T matrix, B be an m × T matrix such that BQ = 0, and ε be distributed as $N(0,\sigma^2 I_T)$. Then $B\varepsilon$ and $\varepsilon'Q\varepsilon$ are distributed independently.

Proof: As in the proof of Lemma 2, let P be the orthogonal matrix such that

$$P'QP = \begin{bmatrix} I_* & 0 \\ 0 & 0 \end{bmatrix}$$

and define $v = P'\varepsilon$; recall that v then has distribution $N(0,\sigma^2 I_T)$. Let r = trace Q, the dimension of I_*. Let v be partitioned as

$$v = \begin{bmatrix} v_1 \\ v_2 \end{bmatrix}$$

v_1 and v_2 being of dimensions r and $T - r$ respectively. Note that v_1 and v_2 are independent of each other.

Now

$$\varepsilon'Q\varepsilon = v'\begin{bmatrix} I_* & 0 \\ 0 & 0 \end{bmatrix}v = v_1'v_1$$

which is really the gist of Lemma 2.

Let $BP = C$, and partition C as (C_1,C_2), where C_1 is of dimension $m \times r$ and C_2 of dimension $m \times (T - r)$. Then

$$C(P'QP) = BPP'QP = BQP = 0 \quad \text{(since } BQ = 0\text{)}$$

Therefore,

$$(C_1,C_2)\begin{bmatrix} I_* & 0 \\ 0 & 0 \end{bmatrix} = 0$$

which implies $C_1 = 0$ and $BP = C = (0,C_2)$.

Finally, then, note that

$$B\varepsilon = BPP'\varepsilon = Cv = C_2v_2 \quad \text{(since } C_1 = 0\text{)}$$

Since $B\varepsilon$ depends only on v_2 and $\varepsilon'Q\varepsilon$ depends only on v_1, and since v_1 and v_2 are independent, $B\varepsilon$ and $\varepsilon'Q\varepsilon$ must also be independent.

6. PROPOSITION: $\hat{\beta}$ and s^2 are independent.

Proof: $s^2 = \varepsilon'M\varepsilon/(T - K)$ and $\hat{\beta} - \beta = (X'X)^{-1}X'\varepsilon$. Since $(X'X)^{-1}X'M = 0$, Lemma 5 implies that s^2 is independent of $\hat{\beta} - \beta$ and hence of $\hat{\beta}$.

The results of this subsection can be summarized by the statement that $\hat{\beta}$ and $(T - K)s^2/\sigma^2$ are independently distributed as $N(\beta,\sigma(X'X)^{-1})$ and chi-squared with T - K degrees of freedom respectively.

Efficiency of $\hat{\beta}$ and s^2

7. PROPOSITION: $\hat{\beta}$ and SSE are (jointly) sufficient statistics for β and σ^2.

Proof: The likelihood function of the sample is

$$L = (2\pi)^{-T/2}(\sigma^2)^{-T/2} \exp\left[-\frac{1}{2\sigma^2}(y - X\beta)'(y - X\beta)\right]$$

But

$$\begin{aligned}(y - X\beta)'(y - X\beta) &= y'y - \beta X'y - y'X\beta + \beta'X'X\beta \\ &= y'y - \hat{\beta}'X'y - y'X\hat{\beta} - (\beta - \hat{\beta})'X'y \\ &- y'X(\beta - \hat{\beta}) + \beta'X'X\beta \\ &= [y'y - \hat{\beta}'X'y - y'X\hat{\beta} + \hat{\beta}'X'X\hat{\beta}] \\ &+ [2(\hat{\beta} - \beta)X'X\hat{\beta} - \hat{\beta}'X'X\hat{\beta} + \beta'X'X\beta] \\ &\qquad \text{(using the fact that } X'y = X'X\hat{\beta}) \\ &= SSE + (\hat{\beta} - \beta)'X'X(\hat{\beta} - \beta)\end{aligned}$$

Thus the likelihood function depends on y only through the values of $\hat{\beta}$ and SSE. Hence $\hat{\beta}$ and SSE are sufficient for β and σ^2 by the Neyman-Fisher factorization criterion [see Note 2].

8. THEOREM: $\hat{\beta}$ and s^2 are efficient.

Proof: $\hat{\beta}$ and s^2 are unbiased estimators of β and σ^2, and they are functions of the sufficient statistics $\hat{\beta}$ and SSE. They are therefore efficient according to the Blackwell-Rao theorem [see Note 3].

The efficiency of $\hat{\beta}$ can also be shown through the Cramér-Rao matrix of lower bounds.

9. LEMMA: The multivariate normal density of ε (with parameters β and σ^2) is regular with respect to its first derivatives [see Note 4].

Proof: The logarithmic likelihood function is

$$\log L = -\frac{T}{2}\log(2\pi) - \frac{T}{2}\log(\sigma^2) - \frac{1}{2\sigma^2}(y - X\beta)'(y - X\beta)$$

[see Note 5]

Therefore

$$\frac{\partial \log L}{\partial \beta} = -\frac{1}{\sigma^2}X'(y - X\beta) \qquad E\left(\frac{\partial \log L}{\partial \beta}\right) = 0$$

$$\frac{\partial \log L}{\partial \sigma^2} = -\frac{T}{2\sigma^2} + \frac{1}{2\sigma^4}(y - X\beta)'(y - X\beta)$$

$$E\left(\frac{\partial \log L}{\partial \sigma^2}\right) = -\frac{T}{2\sigma^2} + \frac{1}{2\sigma^4}\cdot T\sigma^2 = 0$$

Letting $\theta = \begin{bmatrix} \beta \\ \sigma^2 \end{bmatrix}$, the regularity condition $E\left(\frac{\partial \log L}{\partial \theta}\right) = 0$

is therefore satisfied.

10. PROPOSITION: The Cramér-Rao lower bounds for the variance of unbiased estimators of β and σ^2 are $\sigma^2(X'X)^{-1}$ and $2\sigma^4/T$, respectively [see Note 6].

Proof: To form the information matrix we continue differentiating log L:

$$\frac{\partial^2 \log L}{\partial \beta\, \partial \beta'} = -\frac{1}{\sigma^2}X'X \qquad -E\left(\frac{\partial^2 \log L}{\partial \beta\, \partial \beta'}\right) = \frac{X'X}{\sigma^2}$$

$$\frac{\partial^2 \log L}{\partial \sigma^4} = \frac{T}{2\sigma^4} - \frac{1}{\sigma^6}(y - X\beta)'(y - X\beta)$$

$$-E\left(\frac{\partial^2 \log L}{\partial \sigma^4}\right) = \frac{T}{2\sigma^4}$$

$$\frac{\partial^2 \log L}{\partial \beta\, \partial \sigma^2} = -\frac{1}{\sigma^4}X'(y - X\beta) \qquad -E\left(\frac{\partial^2 \log L}{\partial \beta\, \partial \sigma^2}\right) = 0$$

Hence the information matrix is

$$I = \begin{bmatrix} \frac{X'X}{\sigma^2} & 0 \\ 0 & \frac{T}{2\sigma^4} \end{bmatrix}$$

Because the density of L is regular (as verified in Lemma 9), the Cramér-Rao lower bounds are the diagonal elements of I^{-1}, which are clearly $\sigma^2(X'X)^{-1}$ and $2\sigma^4/T$.

The variance of $\hat{\beta}$ equals the Cramér-Rao lower bound $\sigma^2(X'X)^{-1}$; this again shows the efficiency of $\hat{\beta}$: The variance of s^2 exceeds the lower bound $2\sigma^4/T$, but s^2 is still efficient.

Asymptotic Efficiency of $\hat{\beta}$ and s^2

11. PROPOSITION: The maximum likelihood estimators of β and σ^2 are

$$\hat{\beta} = (X'X)^{-1}X'y \qquad \text{and} \qquad \hat{\sigma}^2 = \frac{1}{T}\,\text{SSE}$$

Proof: To maximize L, we set the first derivatives of log L equal to zero.

$$\frac{1}{\hat{\sigma}^2} X'(y - X\hat{\beta}) = 0$$

$$-\frac{T}{2\hat{\sigma}^2} + \frac{1}{2\hat{\sigma}^4}(y - X\hat{\beta})'(y - X\hat{\beta}) = 0$$

Solving the first equation for $\hat{\beta}$ we get

$$\hat{\beta} = (X'X)^{-1}X'y$$

The second equation yields

$$\hat{\sigma}^2 = \frac{1}{T}(y - X\hat{\beta})'(y - X\hat{\beta}) = \frac{1}{T}\,\text{SSE}$$

Since maximum likelihood estimators are asymptotically efficient (under general conditions), one could proceed to demonstrate the asymptotic efficiency of $\hat{\beta}$ and s^2 in this way. However, we will examine the asymptotic efficiency of $\hat{\beta}$ and s^2 through the Cramér-Rao bounds, since the regularity condition for these bounds has already been verified.

12. PROPOSITION: The asymptotic distribution of $\sqrt{T}(\hat{\beta} - \beta)$ is $N(0,\sigma^2\underset{\sim}{Q}^{-1})$, where $\underset{\sim}{Q} = \lim_{T\to\infty}(X'X/T)$.

Proof: For *any* sample size T, the distribution of $\sqrt{T}(\hat{\beta} - \beta)$ is $N(0,\sigma^2(X'X/T)^{-1})$. The above limiting result is therefore trivial [see Note 7].

13. PROPOSITION: $\hat{\beta}$ is asymptotically efficient.

Proof: From Proposition 12, it is clear that the asymptotic variance of $\hat{\beta}$ is

$$\frac{\sigma^2}{T} \underset{\sim}{Q}^{-1}$$

On the other hand, the Cramér-Rao lower bound is the appropriate element of

$$\frac{1}{T} \lim_{T\to\infty} \left(\frac{I}{T}\right)^{-1}$$

which is

$$\frac{1}{T} \lim_{T\to\infty} \sigma^2 \left(\frac{X'X}{T}\right)^{-1} = \frac{\sigma^2}{T} \underset{\sim}{Q}^{-1}$$

Hence the asymptotic variance of $\hat{\beta}$ equals the Cramér-Rao lower bound.

14. PROPOSITION: The asymptotic distribution of $\sqrt{T}(s^2 - \sigma^2)$ is $N(0,2\sigma^4)$.

Proof: By Theorem 3, the distribution of SSE/σ^2 is chi-squared with T - K degrees of freedom. Therefore

$$\frac{SSE}{\sigma^2} = \sum_{t=1}^{T-K} v_t^2$$

where the v_t are independent N(0,1) variables and the v_t^2 are independent chi-squared variables with one degree of freedom. But this is a sum of independent and identically distributed random variables with mean 1 and variance 2; according to the Lindberg-Levy central limit theorem [see Note 8] it follows that

$$\frac{1}{\sqrt{T - K}} \sum_{t=1}^{T-K} \left(\frac{v_t^2 - 1}{\sqrt{2}}\right) \to N(0,1)$$

But this is equivalent to saying that

$$\frac{1}{\sqrt{T - K}}\left[\frac{SSE}{\sigma^2} - (T - K)\right] \to N(0,2)$$

or that

$$\sqrt{T - K}\,(s^2 - \sigma^2) \to N(0,2\sigma^4)$$

or that

$$\sqrt{T}(s^2 - \sigma^2) \to N(0,2\sigma^4)$$

15. PROPOSITION: s^2 is asymptotically efficient.

Proof: From Proposition 14 the asymptotic variance of s^2 is $2\sigma^4/T$, which equals the Cramér-Rao lower bound.

1.5 TESTS OF HYPOTHESES

This section will assume the full ideal conditions, including normality of the disturbances.

Tests Based on the t Distribution

1. LEMMA: Let R be a $1 \times K$ vector, and define s_* by

$$s_* = \sqrt{s^2R(X'X)^{-1}R'}$$

Then $R(\hat{\beta} - \beta)/s_*$ has a t distribution with T - K degrees of freedom (is distributed as t_{T-K}).

Proof: Clearly $R(\hat{\beta} - \beta)$ is a scalar random variable with zero mean and variance $\sigma^2R(X'X)^{-1}R'$; call this variance σ_*^2. Then

$$\frac{R(\hat{\beta} - \beta)}{\sigma_*}$$

is distributed as N(0,1).

Now, from Theorem 3 of Section 1.4 we know that $(T - K)s^2/\sigma^2$ has a chi-squared distribution with T - K degrees of freedom (is distributed as χ^2_{T-K}). Therefore

$$\frac{s_*^2}{\sigma_*^2} = \frac{s^2}{\sigma^2}$$

is distributed as $\chi^2_{T-K}/(T - K)$, a chi-squared variable divided by its number of degrees of freedom.

Finally, then,

$$\frac{R(\hat{\beta} - \beta)}{s_*} = \frac{R(\hat{\beta} - \beta)/\sigma_*}{\sqrt{s_*^2/\sigma_*^2}}$$

is the ratio of a N(0,1) variable to the square root of a $\chi^2_{T-K}/(T - K)$ variable; since the numerator and denominator are independent (by Proposition 6 of Section 1.4), this ratio is distributed as t_{T-K}.

2. THEOREM (Test of a Single Linear Restriction on β): Let R be a known $1 \times K$ vector, and r a known scalar. Define s_* as in Lemma 1. Then under the null hypothesis that $R\beta = r$, the test statistic

$$\frac{R\hat{\beta} - r}{s_*}$$

is distributed as t_{T-K}.

Proof: Under the null hypothesis,

$$\frac{R\hat{\beta} - r}{s_*} = \frac{R\hat{\beta} - R\beta}{s_*} = \frac{R(\hat{\beta} - \beta)}{s_*}$$

so that Lemma 1 is applicable.

It will be shown in the next subsection that the above test is in fact the likelihood ratio test of the hypothesis $R\beta = r$.

3. COROLLARY (Test of Significance of β_i): Let

$$s_{\hat{\beta}_i} = \sqrt{s^2(X'X)^{-1}_{ii}}$$

Then under the null hypothesis that $\beta_i = 0$, the test statistic

$$\frac{\hat{\beta}_i}{s_{\hat{\beta}_i}}$$

is distributed as t_{T-K}.

Proof: This is a special case of Theorem 2, with $r = 0$ and R being a vector of zeroes except for a one in the i-th position.

It is interesting (though of no particular importance, in this context) that there is a simple relationship between the t statistic of Corollary 3 and the (empirical) partial correlation coefficient between y and the i-th regressor, as long as one of the other regressors is a constant term. To develop this relationship, we may as well consider the first regressor; accordingly, let us partition X as

$$X = (x, w)$$

where x is the first regressor and w consists of the last K - 1 regressors.

4. LEMMA: Let $N = I - w(w'w)^{-1}w'$. Then

$$(X'X)^{-1} = \frac{1}{x'Nx}\begin{bmatrix} 1 & -x'w(w'w)^{-1} \\ -(w'w)^{-1}w'x & x'Nx(w'w)^{-1} + (w'w)^{-1}w'xx'w(w'w)^{-1} \end{bmatrix}$$

Proof: The lemma can be checked by multiplying the above matri by

$$X'X = \begin{bmatrix} x'x & x'w \\ w'x & w'w \end{bmatrix}$$

5. LEMMA: Let $M = I - X(X'X)^{-1}X'$. Then

$$M = N - \frac{1}{x'Nx}(Nxx'N)$$

Proof: The result follows after substituting (x,w) for X and the expression in Lemma 4 for $(X'X)^{-1}$.

6. LEMMA: $\hat{\beta}_1 = \dfrac{x'Ny}{x'Nx}$

Proof: $\hat{\beta}_1$ is the first element of $\hat{\beta} = (X'X)^{-1}X'y$. It is ther fore the first row of $(X'X)^{-1}$ times $X'y$, that is,

$$\hat{\beta}_1 = \frac{1}{x'Nx}[1 \quad -x'w(w'w)^{-1}]\begin{bmatrix} x'y \\ w'y \end{bmatrix}$$

$$= \frac{1}{x'Nx} x'[I - w(w'w)^{-1}w']y$$

$$= \frac{x'Ny}{x'Nx}$$

7. LEMMA: If one of the regressors in w is a constant term, then the partial correlation coefficient between y and x is

$$r_{xy\cdot w} = \frac{x'Ny}{\sqrt{x'Nxy'Ny}}$$

Proof: The partial correlation coefficient between x and y is, by definition, the simple correlation coefficient between the two sets of residuals one gets by regressing x and y on w. That is, it is the simple correlation between Nx and Ny. Since these residuals sum to zero if one regressor in w is a constant term, we have

$$r_{xy\cdot w} = \frac{(Nx)'(Ny)}{\sqrt{(Nx)'(Nx)(Ny)'(Ny)}}$$

$$= \frac{x'Ny}{\sqrt{x'Nxy'Ny}}$$

8. LEMMA: $s^2 = \frac{1}{T - K}\left[\frac{x'Nx\ y'Ny - (x'Ny)^2}{x'Nx}\right]$

Proof:

$$s^2 = \frac{1}{T - K} y'My$$

$$= \frac{1}{T - K} y'\left[N - \frac{1}{x'Nx}(Nxx'N)\right]y$$

$$= \frac{1}{T - K}\left(y'Ny - \frac{1}{x'Nx} y'Nx\ x'Ny\right)$$

$$= \frac{1}{T - K}\left(\frac{x'Nxy'Ny - y'Nx\ x'Ny}{x'Nx}\right)$$

9. THEOREM: Let $r_{xy\cdot w}$ be the partial correlation coefficient between y and the first regressor (x). w represents the other K - 1 regressors, one of which is assumed to be a constant term. Let t be the t statistic used to test the hypothesis $\beta_1 = 0$. Then

$$r_{xy\cdot w} = \frac{t}{\sqrt{t^2 + T - K}}$$

Proof:

$$t = \frac{\hat{\beta}_1}{\sqrt{s^2(X'X)^{-1}_{11}}}$$

$$= \frac{\dfrac{x'Ny}{x'Nx}}{\sqrt{\dfrac{1}{T - K} \cdot \dfrac{x'Nxy'Ny - (x'Ny)^2}{x'Nx} \cdot \dfrac{1}{x'Nx}}} \quad \text{(using Lemmas 4,6 and 8)}$$

so that

$$t^2 = \frac{(T - K)(x'Ny)^2}{x'Nxy'Ny - (x'Ny)^2}$$

Dividing both numerator and denominator by $x'Nxy'Ny$ yields

$$t^2 = \frac{(T - K)r^2}{1 - r^2}$$

where r is short for $r_{xy\cdot w}$, or

$$r^2 = \frac{t^2}{t^2 + T - K}$$

The result then follows by taking the square root of both sides of the equation, noting that the sign of r must be the same as the sign of t (since both have the sign of $x'Ny$).

Tests Based on the F Distribution

10. THEOREM (Test of Several Linear Restrictions on β): Let R be a known matrix of dimension $m \times K$ and rank m, r a known $m \times 1$ vector, and $\hat{\beta}$ the OLS estimator. Then under the null hypothesis that $R\beta = r$, the test statistic

$$\frac{(r - R\hat{\beta})'[R(X'X)^{-1}R']^{-1}(r - R\hat{\beta})/m}{SSE/(T - K)}$$

has an F distribution with m and T - K degrees of freedom (is distributed as $F_{m,T-K}$).

Proof: If $r = R\beta$, $r - R\hat{\beta} = R(\beta - \hat{\beta}) = -R(X'X)^{-1}X'\varepsilon$. Therefore the quadratic form in the numerator of the test statistic can be written as $\varepsilon'Q\varepsilon$, where

$$Q = X(X'X)^{-1}R'[R(X'X)^{-1}R']^{-1}R(X'X)^{-1}X'$$

Clearly Q is symmetric and idempotent. Also

$$\begin{aligned} \text{trace } Q &= \text{trace } X(X'X)^{-1}R'[R(X'X)^{-1}R']^{-1}R(X'X)^{-1}X' \\ &= \text{trace}[R(X'X)^{-1}R']^{-1}R(X'X)^{-1}X'X(X'X)^{-1}R' \\ &= \text{trace } I_m = m \end{aligned}$$

Hence the numerator of the above test statistic is distributed as $\sigma^2(\chi^2_m/m)$ by Lemma 2 of Section 1.4.

On the other hand, the denominator is just $\varepsilon'M\varepsilon/(T - K)$, where $M = I - X(X'X)^{-1}X'$ is symmetric, idempotent, and has trace T - K. Hence the denominator is distributed as $\sigma^2\chi^2_{T-K}/(T - K)$.

Finally, the numerator and denominator are independent by Lemma 5 of Section 1.4 since

$$MX(X'X)^{-1}R'[R(X'X)^{-1}R']^{-1}R(X'X)^{-1}X' = 0 \qquad (\text{since } MX = 0)$$

Hence the test statistic is the ratio of two independent chi-squared variables divided by their respective degrees of freedom; it therefore has the F distribution claimed.

Clearly the appropriate rejection region is the upper tail of the F distribution. That is, rejection should be based on large deviations of $R\hat{\beta}$ from r, and hence on large values of the test statistic.

The t test of the preceding subsection is of course just the special case of this test corresponding to m = 1; the formal correspondence is that the square of a random variable with distribution t_{T-K} has distribution $F_{1,T-K}$.

11. LEMMA: The estimator of β which maximizes the likelihood function subject to the constraint $R\beta = r$ is $\tilde{\beta}$ defined by

$$\tilde{\beta} = \hat{\beta} + (X'X)^{-1}R'[R(X'X)^{-1}R']^{-1}(r - R\hat{\beta})$$

$\hat{\beta}$ being the OLS estimator. The maximum likelihood estimator of σ^2 is $\tilde{\sigma}^2$ defined by

$$\tilde{\sigma}^2 = \frac{1}{T}(y - X\tilde{\beta})'(y - X\tilde{\beta})$$

Proof: The log likelihood function is

$$\log L = -\frac{T}{2}\log(2\pi) - \frac{T}{2}\log(\sigma^2) - \frac{1}{2\sigma^2}(y - X\beta)'(y - X\beta)$$

Since there is no constraint on σ^2, we can differentiate to get

$$\frac{\partial \log L}{\partial \sigma^2} = -\frac{T}{2\tilde{\sigma}^2} + \frac{1}{2\tilde{\sigma}^4}(y - X\tilde{\beta}) = 0$$

$$\tilde{\sigma}^2 = \frac{1}{T}(y - X\tilde{\beta})'(y - X\tilde{\beta})$$

Substituting this back into the log likelihood function, we get

$$\log L = -\frac{T}{2}\log(2\pi) - \frac{T}{2}\log[\frac{1}{T}(y - X\tilde{\beta})'(y - X\tilde{\beta})] - \frac{T}{2}$$

which is to be maximized with respect to $\tilde{\beta}$ subject to the constraint $R\tilde{\beta} = r$.

Clearly we maximize the likelihood function above by minimizing the sum of squared errors $(y - X\tilde{\beta})'(y - X\tilde{\beta})$. To minimize this subject to the constraint, we form the Lagrangian

$$y'y - 2\tilde{\beta}'X'y + \tilde{\beta}'X'X\tilde{\beta} - \lambda'(r - R\tilde{\beta})$$

λ being an $m \times 1$ vector of Lagrangian multipliers. Differentiating with respect to $\tilde{\beta}$ and λ yields the conditions

$$-2X'y + 2X'X\tilde{\beta} + R'\lambda = 0$$
$$R\tilde{\beta} - r = 0$$

Premultiplying the top equation by $R(X'X)^{-1}$ gives

$$-2R(X'X)^{-1}X'y + 2R\tilde{\beta} + R(X'X)^{-1}R'\lambda = 0$$

Solving for λ, we get

$$\lambda = [R(X'X)^{-1}R']^{-1}[2R\hat{\beta} - 2R\tilde{\beta}] \qquad [\text{since } (X'X)^{-1}X'y = \hat{\beta}]$$

$$= -2[R(X'X)^{-1}R']^{-1}(r - R\hat{\beta}) \qquad [\text{since } R\tilde{\beta} = r]$$

Substituting this back into the top of the two above first-order conditions, one obtains

$$-2X'y + 2X'X\tilde{\beta} - 2R'[R(X'X)^{-1}R']^{-1}(r - R\hat{\beta}) = 0$$

Solving for $\tilde{\beta}$ yields

$$\tilde{\beta} = (X'X)^{-1}X'y + (X'X)^{-1}R'[R(X'X)^{-1}R']^{-1}(r - R\hat{\beta})$$

which is the desired result.

12. THEOREM: The test given in Theorem 10 is the likelihood ratio test of the hypothesis $R\beta = r$ against the alternative $R\beta \neq r$.

Proof: The likelihood function is

$$L = (2\pi)^{-T/2}(\sigma^2)^{-T/2}\exp\left[-\frac{1}{2\sigma^2}(y - X\beta)'(y - X\beta)\right]$$

Under the alternative (unrestricted) hypothesis, the maximum occurs at

$$\hat{\beta} = (X'X)^{-1}X'y$$

$$\hat{\sigma}^2 = \frac{1}{T}(y - X\hat{\beta})'(y - X\hat{\beta})$$

Hence under the alternative hypothesis, the maximized value of L is

$$(2\pi)^{-T/2}\left[\frac{1}{T}(y - X\hat{\beta})'(y - X\hat{\beta})\right]^{-T/2}\exp\left(-\frac{T}{2}\right)$$

On the other hand, the restricted maximum value of L is

$$(2\pi)^{-T/2}\left[\frac{1}{T}(y - X\tilde{\beta})'(y - X\tilde{\beta})\right]^{-T/2}\exp\left(-\frac{T}{2}\right)$$

Therefore the likelihood ratio test calls for rejection of the null hypothesis for large values of

$$\frac{(2\pi)^{-T/2}[(1/T)(y - X\hat{\beta}')(y - X\hat{\beta})]^{-T/2}\exp(-T/2)}{(2\pi)^{-T/2}[(1/T)(y - X\tilde{\beta}')(y - X\tilde{\beta})]^{-T/2}\exp(-T/2)}$$

and hence for large values of

$$\frac{(y - X\tilde{\beta})'(y - X\tilde{\beta})}{(y - X\hat{\beta})'(y - X\hat{\beta})}$$

But

$$\begin{aligned}(y - X\tilde{\beta})'(y - X\tilde{\beta}) &= y'y - 2\tilde{\beta}'X'y + \tilde{\beta}'X'X\tilde{\beta} \\ &= y'y - 2\hat{\beta}'X'y + \hat{\beta}'X'X\hat{\beta} + 2(\hat{\beta} - \tilde{\beta})X'y \\ &+ \tilde{\beta}'X'X\tilde{\beta} - \hat{\beta}'X'X\hat{\beta} \\ &= (y - X\hat{\beta})'(y - X\hat{\beta}) + (\hat{\beta} - \tilde{\beta})'X'X(\hat{\beta} - \tilde{\beta}) \\ &\qquad \text{(using the fact } X'y = X'X\hat{\beta})\end{aligned}$$

Therefore the likelihood ratio test will have a rejection region consisting of large values of

$$\frac{(\hat{\beta} - \tilde{\beta})'X'X(\hat{\beta} - \tilde{\beta})}{(y - X\hat{\beta})'(y - X\hat{\beta})}$$

Finally, since $\tilde{\beta} - \hat{\beta} = (X'X)^{-1}R'[R(X'X)^{-1}R']^{-1}(r - R\hat{\beta})$, the numerator of this fraction is

$$\begin{aligned}&(r - R\hat{\beta})'[R(X'X)^{-1}R']^{-1}R(X'X)^{-1}X'X(X'X)^{-1}R'[R(X'X)^{-1}R']^{-1}(r - R\hat{\beta}) \\ &\qquad = (r - R\hat{\beta})'[R(X'X)^{-1}R']^{-1}(r - R\hat{\beta})\end{aligned}$$

Therefore the likelihood ratio test will have rejection region consisting of large values of

$$\frac{(r - R\hat{\beta})'[R(X'X)^{-1}R']^{-1}(r - R\hat{\beta})}{(y - X\hat{\beta})'(y - X\hat{\beta})}$$

But except for a factor of proportionality this is just the test statistic of Theorem 10.

A number of standard tests can be obtained as special cases of Theorem 10.

13. COROLLARY (Test of Equality Restriction on β): Under the null hypothesis that $\beta = \beta^*$, the test statistic

$$\frac{(\hat{\beta} - \beta^*)'X'X(\hat{\beta} - \beta^*)/K}{SSE/(T - K)}$$

has an F distribution with K and T - K degrees of freedom.

Proof: This is the special case of Theorem 10 with $m = K$, $R = I_K$, and $r = \beta^*$.

14. LEMMA: Let $X = (X_1, X_2)$, $D = X_2'X_2 - X_2'X_1(X_1'X_1)^{-1}X_1'X_2$. Then

$$(X'X)^{-1} = \begin{bmatrix} (X_1'X_1)^{-1}[I + X_1'X_2D^{-1}X_2'X_1(X_1'X_1)^{-1}] & -(X_1'X_1)^{-1}X_1'X_2D^{-1} \\ -D^{-1}X_2'X_1(X_1X_1)^{-1} & D^{-1} \end{bmatrix}$$

Proof: If E is nonsingular, then

$$\begin{bmatrix} E & F \\ G & H \end{bmatrix}^{-1} = \begin{bmatrix} E^{-1}(I + FD^{-1}GE^{-1}) & -E^{-1}FD^{-1} \\ -D^{-1}GE^{-1} & D^{-1} \end{bmatrix}$$

where $D = H - GE^{-1}F$. (This is sometimes referred to as the partitioned inverse rule.) But the above lemma is just an application of this rule with $E = X_1'X_1$, $F = X_1'X_2$, $G = X_2'X_1$, and $H = X_2'X_2$.

15. PROPOSITION (Test of Equality Restrictions on Several Elements of β): Let the regression equation be arranged as

$$y = X_1\beta_1 + X_2\beta_2 + \epsilon$$

β_1 and β_2 being of dimensions H and K - H, respectively. Let

$\hat{\beta} = \begin{bmatrix} \hat{\beta}_1 \\ \hat{\beta}_2 \end{bmatrix}$, and let β_2^* be a known K - H vector. Let

$D = X_2'[I - X_1(X_1'X_1)^{-1}X_1']X_2 = X_2'M_1X_2$ (as in Lemma 14). Then under the null hypothesis that $\beta_2 = \beta_2^*$, the test statistic

$$\frac{(\hat{\beta}_2 - \beta_2^*)'D(\hat{\beta}_2 - \beta_2^*)/(K - H)}{SSE/(T - K)}$$

i- distributed as $F_{K-H,T-K}$.

Proof: The restriction being tested is of the form $R\beta = r$, where $R = (0, I_{K-H})$ and $r = \beta_2^*$. Then

$$[R(X'X)^{-1}R']^{-1} = [D^{-1}]^{-1} = D \qquad \text{(using Lemma 14)}$$

Also $r - R\hat{\beta} = \beta_2^* - \hat{\beta}_2$. Therefore the proposition follows from Theorem 10.

16. PROPOSITION: Let the regression equation again be arranged as

$$y = X_1\beta_1 + X_2\beta_2 + \varepsilon$$

β_1 and β_2 being of dimensions H and K - H, respectively. Let SSE_H be the sum of squared errors when y is regressed on X_1 only, and SSE_K the sum of squared errors when y is regressed on $X = (X_1, X_2)$. Then under the null hypothesis that $\beta_2 = 0$, the test statistic

$$\frac{(SSE_H - SSE_K)/(K - H)}{SSE_K/(T - K)}$$

is distributed as $F_{K-H,T-K}$.

Proof: Proposition 16 is really just a special case of Proposition 15; however, a direct proof is probably easier than to apply Proposition 15.

Note that

$$\begin{aligned} SSE_H - SSE_K &= y'M_1y - y'My \qquad [\text{where } M_1 = I - X_1'(X_1'X_1)^{-1}X_1'] \\ &= \varepsilon'M_1\varepsilon + \beta_2'X_2'M_1X_2\beta_2 - \varepsilon'M\varepsilon + 2\varepsilon'M_1X_2\beta_2 \\ &= \varepsilon'(M_1 - M)\varepsilon \qquad (\text{if } \beta_2 = 0) \end{aligned}$$

Clearly $M_1 - M$ is symmetric. Also $(M_1 - M)^2 = M_1^2 - M_1M - MM_1 + M^2 = M_1 - 2M_1M + M$; it is therefore idempotent if $M_1M = M$. To show that this is the case, note that

$$M_1M = M - X_1(X_1'X_1)^{-1}X_1'M$$

But since $X'M = 0$, certainly $X_1'M = 0$; hence $M_1M = M$, and $M_1 - M$ is indeed idempotent. Since $M_1 - M$ has trace equal to $(T - H) - (T - K) = K - H$, it follows that $(SSE_H - SSE_K)/(K - H)$ is distributed as $\sigma^2\chi^2_{K-H}/(K - H)$.

We also know that $SSE_K/(T - K)$ is distributed as $\sigma^2\chi^2_{T-K}/(T - K)$. It is also true that the numerator and denominator of the test statistic are independent, since

$$(M_1 - M)M = M - M = 0$$

Hence the test statistic is the ratio of two chi-squared variables divided by their respective number of degrees of freedom, which implies that it has the F distribution claimed.

17. COMMENT (Test of the Existence of a Relationship): If the first regressor is a constant term, one often "tests the existence of a relationship" between y and $X_2, \cdots, X_K$ by testing the hypothesis $\beta_2 = \beta_3 = \cdots = \beta_K = 0$. This can clearly be done through the above test. But note that since $SSE_1 = SST$ (*if* the one regressor "left in" is a constant term), then $SSE_1 - SSE_K = SST - SSE = SSR$. Thus the test statistic is just

$$\frac{SSR/(K - 1)}{SSE/(T - K)}$$

Alternatively, since $SSE = (1 - R^2)SST$ and $SSR = R^2SST$, the test statistic is just

$$\frac{R^2/(K - 1)}{(1 - R^2)/(T - K)}$$

which then is distributed as $F_{K-1,T-K}$ under the hypothesis that $\beta_2 = \beta_3 = \cdots = \beta_K = 0$.

18. THEOREM (Test of Equality of Two Regression Populations): Suppose that one has T_1 observations on a regression equation

$$y_1 = X_1\beta_1 + \varepsilon_1$$

and T_2 observations on another regression equation

$$y_2 = X_2\beta_2 + \varepsilon_2$$

Suppose that X_1 and X_2 are both made up of K regressors. Let SSE_1 denote the sum of squared errors in the regression of y_1 on X_1 and

SSE_2 denote the sum of squared errors in the regression of y_2 on X_2. Finally, let the "joint regression" equation be

$$\begin{bmatrix} y_1 \\ y_2 \end{bmatrix} = \begin{bmatrix} X_1 \\ X_2 \end{bmatrix}\beta + \begin{bmatrix} \varepsilon_1 \\ \varepsilon_2 \end{bmatrix}$$

or simply $y = X\beta + \varepsilon$; let $T_1 + T_2 = T$ be the total number of observations and SSE be the sum of squared errors in the joint regression. Then under the null hypothesis that $\beta_1 = \beta_2$ *and* that

$\varepsilon = \begin{bmatrix} \varepsilon_1 \\ \varepsilon_2 \end{bmatrix}$ is distributed as $N(0,\sigma^2 I_T)$, the test statistic

$$\frac{(SSE - SSE_1 - SSE_2)/K}{(SSE_1 + SSE_2)/(T - 2K)}$$

is distributed as $F_{K,T-2K}$.

Proof: Once again it would be possible to use Theorem 10, but a direct proof is perhaps simpler.

Let $X^* = \begin{bmatrix} X_1 & 0 \\ 0 & X_2 \end{bmatrix}$, $M^* = I - X^*(X^{*\prime}X^*)^{-1}X^{*\prime}$, and SSE^* be the sum of squared errors in the regression of y on X^*. Since the equation

$$y = X^*\begin{bmatrix} \beta_1 \\ \beta_2 \end{bmatrix} + \varepsilon$$

satisfies the full ideal conditions (under the null hypothesis), we know that SSE^* is distributed as $\sigma^2\chi^2_{T-2K}$. But $SSE^* = \varepsilon'M^*\varepsilon = \varepsilon'M_1\varepsilon + \varepsilon'M_2\varepsilon = SSE_1 + SSE_2$ because of the block diagonality of X^*; hence the denominator of the test statistic is distributed as

$$\sigma^2\chi^2_{T-2K}/(T - 2K)$$

Now the quadratic form in the numerator of the test statistic is

$$SSE - SSE^* = y'My - y'M^*y$$
$$= \varepsilon'M\varepsilon - \varepsilon'M^*\varepsilon \quad \text{(under the null hypothesis)}$$

[see Note 1].

But $M - M^*$ is symmetric and has trace equal to $(T - K) - (T - 2K) = K$. To show that it is idempotent, note that

$$(M - M^*)^2 = M^2 - MM^* - M^*M + M^{*2} = M - 2MM^* + M^*$$

so that what we need to show is that $MM^* = M^*$. But $MM^* = M^* - X(X'X)^{-1}X'M^*$. Since $X = X^*\begin{bmatrix} I \\ I \end{bmatrix}$, $X'M^* = (I,I)X^{*\prime}M^* = 0$. Hence $MM^* = M^*$, and $M - M^*$ is idempotent. So the numerator of the above test statistic is distributed as $\sigma^2(\chi_K^2/K)$.

Finally, the numerator and denominator of the statistic are independent, since $M^*(M - M^*) = M^* - M^* = 0$. Hence the test statistic is the ratio of two independent chi-squared variables divided their number of degrees of freedom, and therefore has the F distribution claimed.

EXERCISES

1. Suppose that the first column of the regressor matrix X is a constant term. Define X^* as the last $K - 1$ columns of X, measured in deviations from their respective means; that is,

$$X^*_{ti} = X_{t,i+1} - \frac{1}{T}\sum_{s=1}^{T} X_{s,i+1}$$

   Let y^* be the vector y, measured in deviations from its mean. Show that $\hat{\beta}^* = (X^{*\prime}X^*)^{-1}X^{*\prime}y^*$ is identical to the last $K - 1$ elements of $\hat{\beta} = (X'X)^{-1}X'y$.

2. Find the distribution of the restricted least squares estimator given in Lemma 11 of Section 1.5. Find the distribution of the associated estimator of the error variance, and show that the estimated error variance is distributed independently of the (restricted) coefficient estimator. Explain how you would test a further linear restriction on the coefficients.

3. Show that when one restriction is tested ($m = 1$), the F statistic given in Theorem 10 of Section 1.5 is the square of the t statistic given in Theorem 2 of Section 1.5.
4. Prove Proposition 16 of Section 1.5 as a Corollary of Proposition 15 of Section 1.5.
5. Prove Theorem 18 of Section 1.5 as a Corollary of Theorem 10 of Section 1.5.
6. Verify the "partitioned inversion rule" given in the proof of Lemma 14 of Section 1.5.

NOTES

Section 1.1

1. This term is due to Anscombe and Tukey (1963).

Section 1.2

1. The "hat" over β indicates that this equation holds only at the minimum point. That is, the first derivative here is

$$-2X'y + 2X'X\beta$$

the necessary first-order condition for a minimum is that

$$-2X'y + 2X'X\hat{\beta} = 0$$

2. Sometimes R^2 is defined as SSR/SST. In this case, R^2 will clearly always be nonnegative, but it need not be less than or equal to one unless one of the regressors is a constant term.

Section 1.3

1. The fact that $\text{plim}(X'\varepsilon/T) = 0$ will be used repeatedly. It is perhaps worth noting, therefore, that it is *not* legitimate to claim that $\text{plim}(X'\varepsilon/T) = 0$ simply because X (being nonstochastic) is uncorrelated with ε. For example, if one of the regressors were of the form e^t, this result would not hold; it is this type

of regressor which is ruled out by the requirement that $\lim_{T\to\infty}(X'X/T)$ be finite.

2. Used here are the facts that trace AB = trace BA (if both AB and BA are defined and square), and that trace(A + B) = trace A + trace B. (The trace of a square matrix is of course just the sum of the diagonal elements.)

Section 1.4

1. See the proof of Theorem 11 of Section 1.3.
2. Theorem 8 of Section A.2.
3. Theorem 9 of Section A.2.
4. Regularity is defined in Definition 12 of Section A.2.
5. log(x) denotes the natural logarithm (base e) of X.
6. See Theorem 13 of Section A.2 and the following discussion.
7. See the discussion following Theorem 8 of Section A.1.
8. Theorem 1 of Section A.3.

Section 1.5

1. Note that $X = X^*\begin{bmatrix} I \\ \vdots \\ I \end{bmatrix}$; therefore $M^*X = M^*X^*\begin{bmatrix} I \\ \vdots \\ I \end{bmatrix} = 0$.

REFERENCES

Anscombe, F.J., and J.W. Tukey (1963), "The Examination and Analysis of Residuals," *Technometrics 5*, 141-160.

Chow, G.C. (1960), "Tests for Equality between Sets of Coefficients in Two Linear Regressions," *Econometrica 28*, 591-605.

Fisher, F.M. (1970), "Tests of Equality between Sets of Coefficients of Two Linear Regressions: An Expository Note," *Econometrica 38*, 361-366.

Goldberger, A.S. (1964), *Econometric Theory* New York: John Wiley.

Johnston, J. (1972), *Econometric Methods*, 2nd ed. New York: McGraw Hill.

Quandt, R. (1960), "Tests of the Hypothesis that a Linear Regression System Obeys Two Separate Regimes," *Journal of the American Statistical Association 53,* 324-330.

Theil, H. (1971), *Principles of Econometrics* New York: John Wiley.

Chapter 2

VIOLATIONS OF THE IDEAL CONDITIONS

2.1 INTRODUCTION

The previous chapter dealt with the estimation and hypothesis-testing procedures typically used in a regression model satisfying the full ideal conditions. These conditions were defined in Definition 1 of Section 1.1; they will be restated here in a slightly different manner.

1. COMMENT: The regression model $y = X\beta + \varepsilon$ satisfies the full ideal conditions if:

(i) X is nonstochastic

(ii) The regressors (columns of X) are linearly independent

(iii) $\lim_{T\to\infty} \frac{X'X}{T} = \underset{\sim}{Q}$ is finite and nonsingular

(iv) ε has a multivariate normal distribution

(v) $E(\varepsilon) = 0$

(vi) $E(\varepsilon\varepsilon') = \sigma^2 I$

This chapter will consider violations of the above conditions one at a time, in the sense that it will consider a violation of one of these conditions while assuming that all the others still hold.

Consideration of stochastic regressors, however, will be put off until Chapter 3.

2.2 DISTURBANCES WITH NONZERO MEAN

Suppose that all of the full ideal conditions are satisfied *except* that ε has a nonzero mean vector, say ξ.

1. PROPOSITION: Let $E(\varepsilon) = \xi$. Then

$$E(\hat{\beta}) = \beta + (X'X)^{-1}X'\xi$$

$$\text{plim}\ \hat{\beta} = \beta + \underset{\sim}{Q}^{-1}\lim_{T\to\infty} X'\xi/T$$

Proof: Let $\varepsilon = \xi + V$, where V is then distributed as $N(0,\sigma^2 I)$. Then

$$\hat{\beta} = (X'X)^{-1}X'y = \beta + (X'X)^{-1}X'\xi + (X'X)^{-1}X'V$$

$$E(\hat{\beta}) = \beta + (X'X)^{-1}X'\xi$$

$$\text{plim}\ \hat{\beta} = \beta + \lim\left(\frac{X'X}{T}\right)^{-1} \frac{X'\xi}{T} + \lim_{T\to\infty}\left(\frac{X'X}{T}\right)^{-1} \text{plim}\ \frac{X'V}{T}$$

$$= \beta + \underset{\sim}{Q}^{-1}\lim_{T\to\infty} \frac{X'\xi}{T} \qquad \left(\text{since } \lim_{T\to\infty}\left(\frac{X'X}{T}\right)^{-1} \text{plim}\ \frac{X'V}{T} = 0\right)$$

[see Note 1]

2. COROLLARY: $\hat{\beta}$ is biased unless $X'\xi = 0$, and is inconsistent unless $\lim_{T\to\infty} X'\xi/T = 0$.

3. PROPOSITION: Let $E(\varepsilon) = \xi$ and $M = I - X(X'X)^{-1}X'$. Then

$$E(s^2) = \sigma^2 + \frac{\xi'M\xi}{T-K}$$

$$\text{plim } s^2 = \sigma^2 + \lim_{T\to\infty} \frac{\xi'M\xi}{T}$$

Proof: $s^2 = \frac{\varepsilon'M\varepsilon}{T-K} = \frac{V'MV}{T-K} + \frac{\xi'M\xi}{T-K} + 2\,\frac{V'M\xi}{T-K}$, where as before $\varepsilon = \xi + V$. Therefore,

$$E(s^2) = \sigma^2 + \frac{\xi'M\xi}{T-K} + 2\cdot 0 \qquad \text{[see Note 2]}$$

$$\text{plim } s^2 = \sigma^2 + \lim_{T\to\infty} \frac{\xi'M\xi}{T} + 2 \text{ plim } \frac{V'M\xi}{T}$$

But if $\lim_{T\to\infty} \xi'M\xi/T$ is finite, then necessarily plim $V'M\xi/T = 0$; this is true since $V'M\xi/T$ has zero mean and variance $\sigma^2(\xi'M\xi/T^2)$. Hence in this case it follows that

$$\text{plim } s^2 = \sigma^2 + \lim_{T\to\infty} \frac{\xi'M\xi}{T}$$

4. COROLLARY: s^2 is biased unless $\xi'M\xi = 0$; it is inconsistent unless $\lim_{T\to\infty}(\xi'M\xi/T) = 0$.

5. COMMENT: Since M is a positive semidefinite matrix, $\xi'M\xi \geq 0$. Therefore it is clear that s^2 is biased upward in this case.

6. THEOREM: If $E(\varepsilon) = \xi \neq 0$, either $\hat{\beta}$ or s^2 (or both) *must* be biased. Furthermore, as long as $\lim_{T\to\infty}(\xi'\xi/T) \neq 0$, either $\hat{\beta}$ or s^2 (or both) *must* be inconsistent.

Proof: $\hat{\beta}$ is unbiased only if $X'\xi = 0$. But then

$$\xi'M\xi = \xi'\xi - \xi'X(X'X)^{-1}X'\xi = \xi'\xi > 0$$

as long as $\xi \neq 0$, so that s^2 is biased (according to Corollary 4).

Also $\hat{\beta}$ is inconsistent unless $\lim_{T\to\infty}(X'\xi/T) = 0$; in this case $\lim_{T\to\infty}(\xi'M\xi/T) = \lim_{T\to\infty}(\xi'\xi/T)$, and s^2 is inconsistent unless this is zero.

The above results show that a nonzero mean of the disturbances has serious consequences. This is especially true since the usual statistical tests (such as those of Section 1.5) will all generally be invalid. The usual least squares procedures therefore do not yield estimators of $\hat{\beta}$ and s^2 having the desirable properties of unbiasedness and consistency, nor do they lead to valid tests of hypotheses.

A Special Case

There is, however, a special case in which the situation is not quite so gloomy as pictured above.

7. PROPOSITION: Let the regression equation be partitioned as

$$y = X_1\beta_1 + X_2\beta_2 + \varepsilon$$

Let $\hat{\beta}_1$, $\hat{\beta}_2$, and s^2 be the usual least squares estimators. Now suppose that

$$E(\varepsilon) = \xi = X_1\gamma$$

that is, the mean vector of the disturbances is a linear combination of *some* of the regressors. Then $\hat{\beta}_2$ and s^2 are unbiased and consistent; $E(\hat{\beta}_1) = \text{plim}\ \hat{\beta}_1 = \beta_1 + \gamma$.

Proof: $\hat{\beta} = \beta + (X'X)^{-1}X'\xi + (X'X)^{-1}X'V$. Therefore

$$\begin{aligned} E(\hat{\beta}) &= \beta + (X'X)^{-1}X'\xi \\ &= \beta + (X'X)^{-1}X'X_1\gamma \\ &= \begin{bmatrix} \beta_1 \\ \beta_2 \end{bmatrix} + \begin{bmatrix} X_1'X_1 & X_1'X_2 \\ X_2'X_1 & X_2'X_2 \end{bmatrix}^{-1} \begin{bmatrix} X_1'X_1\gamma \\ X_2'X_1\gamma \end{bmatrix} \\ & \begin{bmatrix} \beta_1 \\ \beta_2 \end{bmatrix} + \begin{bmatrix} I \\ 0 \end{bmatrix}\gamma = \begin{bmatrix} \beta_1 + \gamma \\ \beta_2 \end{bmatrix} \qquad \text{[see Note 3]} \end{aligned}$$

That $\text{plim}\ \hat{\beta} = E(\hat{\beta})$ should be clear since $\text{plim}(X'X)^{-1}X'V = 0$.

Unbiasedness of s^2 follows from the fact that its bias is

$$\xi'M\xi = \gamma'X_1'MX_1\gamma$$

Clearly the fact that $MX = 0$ implies that $MX_1 = 0$, and hence the bias of s^2 is zero. Its consistency should then also be clear.

In general it is not obvious why the mean vector of the disturbances should in any way be related to the regressors, so that the above result is of limited usefulness. However, there is one reasonable case in which it might apply.

8. COROLLARY: Let $E(\varepsilon_t) = \mu$ for all t. Then if one of the regressors is a constant term, s^2 and all of the elements of $\hat{\beta}$ *except* the coefficient of the constant term are unbiased and consistent.

Finally, it may be worth pointing out that these last results do not constitute a counterexample to Proposition 6, since one cannot say that $\hat{\beta}$ is unbiased (or consistent) unless each of its elements is.

Omitted Variables

Suppose one has a regression equation

$$y = X_1\beta_1 + X_2\beta_2 + \varepsilon$$

which satisfies the full ideal conditions. Suppose, however, that y is regressed not on X_1 and X_2 but only on X_1. This amounts to applying least squares to the model

$$y = X_1\beta_1 + \varepsilon^*$$

where $\varepsilon^* = X_2\beta_2 + \varepsilon$. This is called a case of "omitted variables," the variables X_2 having been omitted from the regression equation. Note that $E(\varepsilon^*) = X_2\beta_2$ which is generally not equal to zero; it should therefore be clear that least squares applied to the misspecified equation will not have desirable properties [see Note 4].

9. PROPOSITION: Let the equation

$$y = X_1\beta_1 + X_2\beta_2 + \varepsilon$$

satisfy the full ideal conditions, and suppose that $\hat{\beta}_1$ and s^2 are the estimators obtained by applying least squares to the misspecified equation

$$y = X_1\beta_1 + \varepsilon^*$$

Then

(i) $$E(\hat{\beta}_1) = \beta_1 + (X_1'X_1)^{-1}X_1'X_2\beta_2$$

(ii) $$\operatorname{plim} \hat{\beta}_1 = \lim_{T\to\infty}\left(\frac{X_1'X_1}{T}\right)^{-1} \lim_{T\to\infty}\left(\frac{X_1'X_2}{T}\right) + \beta_1$$

(iii) $$E(s^2) = \sigma^2 + \frac{(X_2\beta_2)'M_1(X_2\beta_2)}{T - K}$$

[where $M_1 = I - X_1(X_1'X_1)^{-1}X_1'$]

(iv) $$\operatorname{plim} s^2 = \sigma^2 + \lim_{T\to\infty} \frac{1}{T}(X_2\beta_2)'M_1(X_2\beta_2)$$

Proof: The above results are just applications of Propositions 1 and 3 with $\xi = X_2\beta_2$.

Note that the only case in which *both* $\hat{\beta}_1$ and s^2 are unbiased or consistent is the case in which $X_2\beta_2 = 0$. Since under the full ideal conditions the columns of X_2 are linearly independent, this can be true only if $\beta_2 = 0$; that is, only if all the omitted variables have zero coefficients.

10. COMMENT: It is interesting, though of no particular importance, to note that

$$E(\hat{\beta}_1) = \beta_1 + R_{12}\beta_2$$

where $R_{12} = (X_1'X_1)^{-1}X_1'X_2$ can be interpreted as the "estimated regression coefficients" in the (descriptive) regression of X_2 on X_1.

2.3 MULTICOLLINEARITY

1. DEFINITION: Multicollinearity is said to exist if the rank of the regressor matrix X is less than K, the number of regressors.

An equivalent definition would of course be to say that multicollinearity exists if the columns of X are linearly dependent.

Estimation of β

2. PROPOSITION: The least squares estimator $\hat{\beta} = (X'X)^{-1}X'y$ does not exist if the rank of X is less than K.

Proof: $X'X$ is of dimension $K \times K$ and has the same rank as X. Hence if the rank of X is less than K, $X'X$ is singular.

3. PROPOSITION: There does not exist an estimator of β which uniquely minimizes SSE.

Proof: Since the columns of X are linearly dependent, there exists a vector γ such that $X\gamma = 0$. Then if $\hat{\beta}$ is an estimator of β, $\tilde{\beta} = \hat{\beta} + \gamma$ is another estimator such that $X\hat{\beta} = X\tilde{\beta}$. Therefore $(y - X\hat{\beta})'(y - X\hat{\beta}) = (y - X\tilde{\beta})'(y - X\tilde{\beta})$; there cannot exist any estimator $\hat{\beta}$ which uniquely minimizes SSE.

4. THEOREM: The vector of parameters β could not be deduced even from knowledge of the likelihood function (given the particular X matrix); it is not "identifiable" [see Note 1].

Proof: The likelihood function is

$$L = (2\pi)^{-T/2}(\sigma^2)^{-T/2}\exp\left[-\frac{1}{2\sigma^2}(y - X\beta)'(y - X\beta)\right]$$

But if γ is a vector such that $X\gamma = 0$, then β and $\beta^* = \beta + \gamma$ yield identical values of the likelihood function since $X\beta = X\beta^*$. The values β and β^* are thus "observationally equivalent" and cannot possibly be distinguished from the likelihood function.

The above results show that it is quite useless to try to estimate β. This is important to realize, since it is possible to solve the set of least squares normal equations

$$X'X\hat{\beta} = X'y$$

for an estimator $\hat{\beta}$. However, this estimator would have none of the usual desirable properties. Indeed, an infinite number of solutions to these equations exist.

One particular way to find a solution to the normal equations is through the use of the concept of "generalized inverse" of a matrix.

5. MATHEMATICAL REVIEW: Let A be any matrix. Then there exists a unique matrix A^+, called the generalized inverse of A, which satisfies the conditions

(i) $AA^+A = A$

(ii) $A^+AA^+ = A^+$

(iii) $(A^+A)' = A^+A$

(iv) $(AA^+)' = AA^+$

The following facts will also be used:

(a) trace AA^+ = rank A = trace A^+A

(b) $b = A^+c$ is a solution to the set of linear equations $Ab = c$.

One particular explicit solution of the set of normal equations

$$X'X\hat{\beta} = X'y$$

is therefore $\hat{\beta} = (X'X)^+X'y$. In some of what follows it will be useful to have this explicit solution available.

6. PROPOSITION: If $\hat{\beta}$ and $\tilde{\beta}$ are any two solutions of the normal equations, then $X\hat{\beta} = X\tilde{\beta}$.

Proof: Since $\hat{\beta}$ and $\tilde{\beta}$ are solutions to the normal equations,

$$X'X(\hat{\beta} - \tilde{\beta}) = 0$$

Therefore,

$$(\hat{\beta} - \tilde{\beta})'X'X(\hat{\beta} - \tilde{\beta}) = 0$$

or

$$[X(\hat{\beta} - \tilde{\beta})]'X(\hat{\beta} - \tilde{\beta}) = 0$$

But this implies that $X(\hat{\beta} - \tilde{\beta}) = 0$.

7. PROPOSITION: Any solution of the normal equations minimizes SSE.

Proof: Let $\hat{\beta}$ be a solution of the normal equations and β be any other value. Then

$$\begin{aligned}(y - X\beta)'(y - X\beta) &= [(y - X\hat{\beta}) + X(\hat{\beta} - \beta)]'[(y - X\hat{\beta}) + X(\hat{\beta} - \beta)] \\ &= (y - X\hat{\beta})'(y - X\hat{\beta}) + (\hat{\beta} - \beta)'X'X(\hat{\beta} - \beta) \\ &\qquad\qquad [\text{since } X'(y - X\hat{\beta}) = 0] \\ &\geq (y - X\hat{\beta})'(y - X\hat{\beta})\end{aligned}$$

8. LEMMA: $X = X(X'X)^{+}X'X$, where $(X'X)^{+}$ is the generalized inverse of $X'X$.

Proof: Let $G = X - X(X'X)^{+}X'X = [I - X(X'X)^{+}X']X$. But then $G'G = X'X - X'X(X'X)^{+}X'X - X'X(X'X)^{+}X'X + X'X(X'X)^{+}X'X(X'X)^{+}X'X = 0$, which implies $G = 0$ and $X = X(X'X)^{+}X'X$.

9. PROPOSITION: Let $\hat{\beta}$ be any solution of the normal equations. Then $X\hat{\beta}$ is an unbiased estimator of $X\beta$.

Proof: Since $X\hat{\beta}$ is invariant with respect to choice of solutions of the normal equations (Proposition 6), we can consider the particular solution $\hat{\beta} = (X'X)^{+}X'y$. Then

$$E(X\hat{\beta}) = X(X'X)^{+}X'X\beta = X\beta$$

since $X(X'X)^{+}X'X = X$ by Lemma 8.

This shows that it is possible to estimate $X\beta$. Of course, it is clear in looking at the likelihood function that $X\beta$, the mean of y, is identifiable. The only problem here is that when X is not of rank K, $X\beta$ does not uniquely determine β.

10. THEOREM (Rao): Let w be a K vector and $w'\beta$ a linear function of β. Then the following are equivalent:

(i) $w'\hat{\beta}$ is linear in y and unbiased (where $\hat{\beta}$ is any solution to the normal equations)

(ii) w is a linear combination of columns of $X'X$

(iii) $w'\hat{\beta}$ is invariant with respect to solutions of the normal equations.

Proof: First we show that (ii) implies (i) and (iii). If w is a linear combination of columns of $X'X$, there exists a vector λ such that $w = X'X\lambda$. Then if $\hat{\beta}$ is any solution to the normal equations

$$w'\hat{\beta} = \lambda'X'X\hat{\beta} = \lambda'X'y$$

which shows (iii) and also that $w'\hat{\beta}$ is linear in y. $w'\hat{\beta}$ is also unbiased since $E(w'\hat{\beta}) = \lambda'X'X\beta = w'\beta$. This shows (i).

Next we show that (iii) implies (ii). We know that there exists a vector γ such that $X\gamma = 0$, and hence $X'X\gamma = 0$. Then if $\hat{\beta}$ is a solution to the normal equations, $\tilde{\beta} = \hat{\beta} - \gamma$ is also a solution. If $w'\hat{\beta}$ is invariant with respect to such solutions, we must have $w'(\hat{\beta} - \tilde{\beta}) = w'\gamma = 0$. But $X'X\gamma = 0$ and $w'\gamma = 0$ imply that w is in the space spanned by the columns of X'X, thus showing (ii).

Finally, we show that (i) implies (ii). If $w'\hat{\beta}$ is linear in y, we can write $w'\hat{\beta} = L'y$. If $w'\hat{\beta}$ is unbiased, $E(w'\hat{\beta}) = E(L'y) = L'X\beta$. Clearly this requires $w' = L'X$ or $w = X'L$, so that w is in the space spanned by the columns of X'. Since the spaces spanned by the columns of X' and by the columns of X'X are in fact the same, this implies that w is a linear combination of columns of X'X, thus showing (ii).

11. DEFINITION: $w'\beta$ is *estimable* if w is a linear combination of columns of X' (or, equivalently, of X'X).

12. COROLLARY: $w'\hat{\beta}$ is invariant with respect to solutions of the normal equations, linear in y, and unbiased for $w'\beta$ if and only if $w'\beta$ is estimable.

All elements of $X\beta$ are clearly estimable according to definition 11, since rows of X *are* columns of X'. This constitutes an alternative proof of Proposition 9.

However, not all linear functions $w'\beta$ are estimable unless X is of rank K.

13. PROPOSITION: If $w'\beta$ is estimable, then the BLUE of $w'\beta$ is $w'\hat{\beta}$, $\hat{\beta}$ being any solution of the normal equations.

Proof: Since $w'\beta$ is estimable, there exists a vector γ such that $w = X'X\gamma$. Therefore $w'\hat{\beta} = \gamma'X'X\hat{\beta} = \gamma'X'y$.

Now let L'y be any other unbiased estimator of $w'\beta$; note that this requires $L'X = w'$. Then

$$\begin{aligned}\mathrm{Var}(L'y) &= \mathrm{Var}(L'y - \gamma'X'y + \gamma'X'y) \\ &= \mathrm{Var}(L'y - \gamma'X'y) + \mathrm{Var}(\gamma'X'y) \\ &+ 2\,\mathrm{Cov}(L'y - \gamma'X'y)(\gamma'X'y)\end{aligned}$$

But

$$\begin{aligned}\mathrm{Cov}(L'y - \gamma'X'y)(\gamma'X'y) &= E(L'\varepsilon - \gamma'X'\varepsilon)(\gamma'X'\varepsilon)' \\ &= \sigma^2(L' - \gamma'X')(X\gamma) \\ &= \sigma^2(L'X - \gamma X'X)\gamma \\ &= \sigma^2(w' - w')\gamma = 0\end{aligned}$$

Therefore

$$\begin{aligned}\mathrm{Var}(L'y) &= \mathrm{Var}(L'y - \gamma'X'y) + \mathrm{Var}(\gamma'X'y)\\ &= \mathrm{Var}(L'y - w'\hat{\beta}) + \mathrm{Var}(w'\hat{\beta})\\ &\geq \mathrm{Var}(w'\hat{\beta})\end{aligned}$$

14. PROPOSITION: If $w'\beta$ is estimable and $\hat{\beta}$ is any solution of the normal equations, then

$$\mathrm{Var}(w'\hat{\beta}) = \sigma^2 w'(X'X)^+ w = \sigma^2\lambda'X'X\lambda$$

λ being any vector such that $w = X'X\lambda$.

Proof: We may as well take $\hat{\beta} = (X'X)^+X'y$. Then $w'\hat{\beta} = w'(X'X)^+X'y$, which has variance

$$\sigma^2 w'(X'X)^+X'X(X'X)^+w = \sigma^2 w'(X'X)^+w$$

Since there exists a λ such that $w = X'X\lambda$ (or else $w'\beta$ would not be estimable), we have

$$\begin{aligned}\mathrm{Var}(w'\hat{\beta}) &= \sigma^2 w'(X'X)^+w\\ &= \sigma^2\lambda'X'X(X'X)^+X'X\lambda\\ &= \sigma^2\lambda'X'X\lambda\end{aligned}$$

Estimation of σ^2

15. PROPOSITION: $E(y - X\hat{\beta}) = 0$, if $\hat{\beta}$ is any solution of the normal equations.

Proof: $E(y - X\hat{\beta}) = X\beta - X\beta = 0$

16. LEMMA: Let $\hat{\beta}$ be any solution of the normal equations. Then $(y - X\hat{\beta})'(y - X\hat{\beta}) = \varepsilon'M_*\varepsilon$, where $M_* = I - X(X'X)^+X'$ is symmetric and idempotent and has trace $M_* = T - \mathrm{rank}(X)$.

Proof: Since $X\hat{\beta}$ is invariant with respect to solutions of the normal equations, we can let $\hat{\beta} = (X'X)^+X'y$. Then $y - X\hat{\beta} = y - X(X'X)^+X'y = M_*y$.

That M_* is symmetric is obvious since $(X'X)^+$ is symmetric. To show that it is idempotent, consider

$$M_*^2 = I - X(X'X)^+X' - X(X'X)^+X' + X(X'X)^+X'X(X'X)^+X' = M_*$$

$$[\text{since } (X'X)^+X'X(X'X)^+ = (X'X)^+]$$

Then

$$(y - X\hat{\beta})'(y - X\hat{\beta}) = (M_*y)'M_*y$$
$$= y'M_*y \quad \text{(since } M_* \text{ is symmetric and idempotent}$$

To show that this equals $\varepsilon'M_*\varepsilon$, all that need be shown is that

$$0 = M_*X = X - X(X'X)^+X'X$$

which is just Lemma 8. Finally,

$$\begin{aligned} \text{trace } M_* &= \text{trace } I - \text{trace } X(X'X)^+X' \\ &= T - \text{trace}(X'X)^+X'X \\ &= T - \text{rank}(X) \end{aligned}$$

since for any matrix A, trace A^+A = rank A.

17. PROPOSITION: Let $\hat{\beta}$ be any solution to the normal equations. Then

$$\frac{(y - X\hat{\beta})'(y - X\hat{\beta})}{\sigma^2}$$

is distributed as χ^2_{T-r}, where r = rank(X).

Proof: The result follows immediately from Lemma 16 and from Lemma 2 of Section 1.4.

18. COROLLARY: An unbiased estimator of σ^2 is s^2 defined by

$$s^2 = \frac{(y - X\hat{\beta})'(y - X\hat{\beta})}{T - \text{rank}(X)}$$

where $\hat{\beta}$ is any solution to the normal equations.

19. PROPOSITION: Let $\hat{\beta}$ be any solution of the normal equations. Then $(y - X\hat{\beta})'(y - X\hat{\beta})$ and $X\hat{\beta}$ are distributed independently.

Proof:

$$X(\hat{\beta} - \beta) = X(X'X)^+X'\varepsilon$$

while

$$(y - X\hat{\beta})'(y - X\hat{\beta}) = \varepsilon'[I - X(X'X)^+X']\varepsilon$$

Since $[I - X(X'X)^+X']X(X'X)^+X' = 0$, the result follows from Lemma 5 of Section 1.4.

It is therefore possible to construct t tests for hypotheses concerning $w'\beta$. The situation is really completely analogous to the usual case in which rank(X) = K.

Use of A Priori Information

It should be clear that the root of the problem here is that there is just not enough information in the data to enable one to make inferences about all K elements of β. Of course a priori information, if available, can be used to identify β and make estimation possible.

Specifically, let m = K - rank(X); the set of normal equations $X'X\beta = X'y$ is only a set of K - m independent equations which cannot be solved uniquely for β. However, suppose that we have m a priori restrictions on β, of the form $R\beta = r$, where R is a known m × K matrix and r is a known m vector. Then the following proposition shows that β can now be estimated.

20. PROPOSITION: Let m = K - rank(X), and suppose that one has m restrictions on β, of the form $R\beta = r$, as above. Suppose that the matrix

$$V = \begin{bmatrix} X'X \\ R \end{bmatrix}$$

has rank K [see Note 2]. Let $\tilde{\beta}$ be the solution to the set of augmented normal equations

$$V\tilde{\beta} = \begin{bmatrix} X'y \\ r \end{bmatrix}$$

Then $\tilde{\beta}$ is an unbiased estimator of β.

Proof: We will use the fact that when V is of full column rank, $V^+ = (V'V)^{-1}V'$. Hence we have

$$\begin{aligned}\tilde{\beta} &= V^+\begin{bmatrix} X'y \\ r \end{bmatrix} \\ &= (X'XX'X + R'R)^{-1}(X'XX'y + R'r) \\ &= (X'XX'X + R'R)^{-1}[(X'XX'X + R'R)\beta + X'XX'\varepsilon] \\ &= \beta + (X'XX'X + R'R)^{-1}X'XX'\varepsilon\end{aligned}$$

Therefore $E(\tilde{\beta}) = \beta$.

Near Multicollinearity

So far in this section we have discussed cases in which rank(X) < K, which means that one of the regressors is a linear combination of the others. This case might be called "perfect multicollinearity." We now turn to the more common case in which one of the regressors is "almost" a linear combination of the others, which may be called "near multicollinearity." Note that in this case X'X is "almost" singular, but it is not singular, so that the OLS estimator can be calculated. We proceed to consider the properties of OLS under near multicollinearity.

21. LEMMA: Let x_j be the j-th column of X, and X_j^* be the other j - 1 columns. Then

$$(X'X)^{-1}_{jj} = \frac{1}{x_j'[I - X_j^*(X_j^{*\prime}X_j^*)^{-1}X_j^{*\prime}]x_j}$$

Proof: Rearrange regressors so that x_j is the first column of X. Then this result is part of Lemma 4 of Section 1.5.

22. COMMENT: The denominator of the right-hand side above is the residual sum of squares when x_j is regressed on X_j^*. If x_j is nearly a linear combination of columns in X_j^*, this sum of squares will be very small. Therefore $(X'X)^{-1}_{jj}$ will be large, which means that the OLS estimate of the j-th element of β will have a large variance. It is often felt that, in the presence of near multicollinearity, the OL estimates are too imprecise to be of much use. This is so despite their being the minimum variance unbiased estimates.

Under these circumstances it is sometimes argued that it may be preferable to consider biased estimates of β, if their variances are sufficiently smaller than those of the OLS estimates. One such biased estimator is the "ridge regression" estimator.

23. DEFINITION: The ridge estimator of β is

$$\tilde{\beta} = (X'X + kI)^{-1}X'y$$

where k > 0 is an arbitrary constant.

24. NOTE: Picking $k = 0$ gives the OLS estimator.

25. PROPOSITION:

$$E(\tilde{\beta}) = (X'X + kI)^{-1}X'X\beta$$

$$\text{Cov}(\tilde{\beta}) = \sigma^2(X'X + kI)^{-1}X'X(X'X + kI)^{-1}$$

26. PROPOSITION: $\text{Cov}(\hat{\beta}) - \text{Cov}(\tilde{\beta})$ is positive semidefinite for any $k > 0$.

Proof: $\text{Cov}(\hat{\beta}) - \text{Cov}(\tilde{\beta})$ is σ^2 times

$$(X'X)^{-1} - (X'X + kI)^{-1}X'X(X'X + kI)^{-1}$$

This is positive semidefinite if and only if

$$(X'X + kI)(X'X)^{-1}(X'X + kI) - (X'X)$$

is positive semidefinite. But simplifying this expression gives

$$2kI + k^2(X'X)^{-1}$$

which is certainly positive semidefinite for $k > 0$.

27. COMMENT: The previous two propositions show that the ridge estimator is biased, but that its variance is less than the OLS estimator's variance. This raises the possibility that its mean square error may be less than the OLS estimator's mean square error.

The comparisons of mean square errors are simplified by looking at the scalar quantity $E(\tilde{\beta} - \beta)'(\tilde{\beta} - \beta)$, which is the trace of the mean square error matrix. It is also the sum of the mean square errors of the estimates of the various elements of β. In doing so, however, we should not fail to recognize that this measure has its defects. In particular, *$E(\tilde{\beta} - \beta)'(\tilde{\beta} - \beta)$ depends on the units of measurement of the explanatory variables.*

28. PROPOSITION: Let $\lambda_1 \cdots \lambda_K$ be the eigenvalues of $X'X$. Then

$$E(\hat{\beta} - \beta)'(\hat{\beta} - \beta) = \sigma^2 \sum_{i=1}^{K} \frac{1}{\lambda_i}$$

Proof:

$$\begin{aligned} E(\hat{\beta} - \beta)'(\hat{\beta} - \beta) &= \text{trace}[\text{MSE}(\hat{\beta})] \\ &= \text{trace}[\text{Cov}(\hat{\beta})] \quad \text{(since } \hat{\beta} \text{ is unbiased)} \\ &= \sigma^2 \,\text{trace}(X'X)^{-1} \\ &= \sigma^2 \sum_{i=1}^{K} \frac{1}{\lambda_i} \end{aligned}$$

29. PROPOSITION:

$$\text{trace}[\text{Cov}(\tilde{\beta})] = \sigma^2 \sum_{i=1}^{K} \frac{\lambda_i}{(\lambda_i + k)^2}$$

Proof:

$$\begin{aligned} \text{trace}[\text{Cov}(\tilde{\beta})] &= \sigma^2 \,\text{trace}[(X'X + kI)^{-1}X'X(X'X + kI)^{-1}] \\ &= \sigma^2 \,\text{trace}[(X'X + kI)^{-1}(X'X + kI)(X'X + kI)^{-1} \\ &\quad - k(X'X + kI)^{-2}] \\ &= \sigma^2 \,\text{trace}(X'X + kI)^{-1} - k\sigma^2 \,\text{trace}(X'X + kI)^{-2} \\ &= \sigma^2 \sum_{i=1}^{K} \frac{1}{\lambda_i + k} - k\sigma^2 \sum_{i=1}^{K} \frac{1}{(\lambda_i + k)^2} \\ &= \sigma^2 \sum_{i=1}^{K} \frac{\lambda_i}{(\lambda_i + k)^2} \end{aligned}$$

30. LEMMA:

$$(X'X + kI)^{-1}X'X = I - k(X'X + kI)^{-1} = [I + k(X'X)^{-1}]^{-1}$$

Proof: We first prove the first equality.

$$\begin{aligned} 0 &= X'X - [X'X + kI] + kI \\ &= (X'X + kI)^{-1}X'X - I + k(X'X + kI)^{-1} \end{aligned}$$

which implies

$$(X'X + kI)^{-1}X'X = I - k(X'X + kI)^{-1}$$

We now prove the second equality. The line immediately above implies

$$(X'X + kI)^{-1}X'X + k(X'X + kI)^{-1} = I$$

or

$$(X'X + kI)^{-1}(X'X)[I + k(X'X)^{-1}] = I$$

which yields

$$(X'X + kI)^{-1}X'X = [I + k(X'X)^{-1}]^{-1}$$

31. LEMMA: $[\text{Bias}(\tilde{\beta})]'[\text{Bias}(\tilde{\beta})] = k^2\beta'(X'X + kI)^{-2}\beta$

Proof: Define $Z = (X'X + kI)^{-1}X'X$, so that $E(\tilde{\beta}) = Z\beta$. Then

$$\begin{aligned}\text{Bias}(\tilde{\beta}) &= E(\tilde{\beta}) - \beta = (Z - I)\beta \\ &= -k(X'X + kI)^{-1}\beta\end{aligned}$$

since, by Lemma 30, $Z = I - k(X'X + kI)^{-1}$. Then clearly

$$[\text{Bias}(\tilde{\beta})]'[\text{Bias}(\tilde{\beta})] = k^2\beta'(X'X + kI)^{-2}\beta$$

32. PROPOSITION: Let P be the orthogonal matrix such that

$$P'(X'X)P = \Lambda$$

where Λ is the diagonal matrix of eigenvalues of $X'X$. (P is the orthogonalized matrix of eigenvectors of $X'X$.) Define

$$\alpha = P'\beta.$$

Then

$$[\text{Bias}(\tilde{\beta})]'[\text{Bias}(\tilde{\beta})] = k^2 \sum_{i=1}^{K} \frac{\alpha_i^2}{(\lambda_i + k)^2}$$

Proof: From Lemma 31,

$$\begin{aligned}[\text{Bias}(\tilde{\beta})]'[\text{Bias}(\tilde{\beta})] &= k^2\beta'(X'X + kI)^{-2}\beta \\ &= k^2\alpha'[P'(X'X + kI)^{-2}P]\alpha\end{aligned}$$

Since $X'X$ and $(X'X + kI)^{-2}$ have the same eigenvectors,

$$P'(X'X + kI)^{-2}P = \psi$$

where ψ is the diagonal matrix of eigenvalues of $(X'X + kI)^{-2}$, of which the i-th is

$$\frac{1}{(\lambda_i + k)^2} \qquad (i = 1, 2, \cdots, K)$$

Therefore,

$$[\text{Bias}(\tilde{\beta})]'[\text{Bias}(\tilde{\beta})] = k^2\alpha'\psi\alpha$$
$$= k^2 \sum_{i=1}^{K} \frac{\alpha_i^2}{(\lambda_i + k)^2}$$

33. THEOREM: Let $\lambda_1 \cdots \lambda_K$ be the eigenvalues of $X'X$, and α as defined in Proposition 32. Then

$$E(\tilde{\beta} - \beta)'(\tilde{\beta} - \beta) = \sigma^2 \sum_{i=1}^{K} \frac{\lambda_i}{(\lambda_i + k)^2} + k^2 \sum_{i=1}^{K} \frac{\alpha_i^2}{(\lambda_i + k)^2}$$

Proof: Obvious from Propositions 29 and 32.

34. PROPOSITION: For $k > 0$,

$$\frac{d}{dk}[\text{Bias}(\tilde{\beta})]'[\text{Bias}(\tilde{\beta})] \geq 0$$
$$\frac{d}{dk}\text{trace}[\text{Cov}(\tilde{\beta})] \leq 0$$

Proof:

$$\frac{d}{dk} k^2 \sum_{i=1}^{K} \frac{\alpha_i^2}{(\lambda_i + k)^2} = 2k \sum_{i=1}^{K} \frac{\alpha_i^2}{(\lambda_i + k)^2} - 2k^2 \sum_{i=1}^{K} \frac{\alpha_i^2}{(\lambda_i + k)^3}$$
$$= 2k \sum_{i=1}^{K} \frac{\alpha_i^2\lambda_i}{(\lambda_i + k)^3} \geq 0$$

$$\frac{d}{dk}\sigma^2 \sum_{i=1}^{K} \frac{\lambda_i}{(\lambda_i + k)^2} = -2\sigma^2 \sum_{i=1}^{K} \frac{\lambda_i}{(\lambda_i + k)^3} \leq 0$$

35. COMMENT: This shows that, the larger k, the larger is the bias of $\tilde{\beta}$, but the smaller is its variance.

36. THEOREM: There exists $k > 0$ such that

$$E(\tilde{\beta} - \beta)'(\tilde{\beta} - \beta) < E(\hat{\beta} - \beta)'(\hat{\beta} - \beta)$$

Proof: We have

$$\frac{d}{dk}E(\tilde{\beta} - \beta)'(\tilde{\beta} - \beta) = 2k\sum_{i=1}^{K}\frac{\alpha_i^2\lambda_i}{(\lambda_i + k)^3} - 2\sigma^2\sum_{i=1}^{K}\frac{\lambda_i}{(\lambda_i + k)^3}$$

which is negative at least for the range

$$0 \leq k < \frac{\sigma^2}{\alpha_{max}{}^2}$$

At this point several comments are in order:

(i) Theorem 36 is not as comforting as might seem, since σ^2 and the α_i are unknown. Picking k based on *estimates* of these quantities leads to an estimator with complicated, and as yet undiscovered, properties.

(ii) The "optimal" value of k could be obtained by setting the derivative of $E(\tilde{\beta} - \beta)'(\tilde{\beta} - \beta)$ to zero and solving for k. Again, this would depend on unknown parameters.

(iii) Recall that the criterion $E(\tilde{\beta} - \beta)'(\tilde{\beta} - \beta)$ depends on the units of measurement of the explanatory variables. However, we can substitute the criterion $E(X\tilde{\beta} - X\beta)'(X\tilde{\beta} - X\beta)$, which is independent of units of measurement, and get similar results.

(iv) To an econometrician, the real point of modeling is often to test hypotheses on β. If this is the case, ridge regression is not of much use and may be misleading.

37. PROPOSITION: For $k > 0$, $\tilde{\beta}'\tilde{\beta} < \hat{\beta}'\hat{\beta}$.

Proof: Note that $\tilde{\beta} = Z\hat{\beta}$, where as before

$$Z = (X'X + kI)^{-1}X'X = [I + k(X'X)^{-1}]^{-1}$$

Therefore,

$$\tilde{\beta}'\tilde{\beta} - \hat{\beta}\,\hat{\beta} = \hat{\beta}'(Z^2 - I)\hat{\beta}$$

Clearly a sufficient condition for this to be ≤ 0 is for $Z^2 - I$ to be negative semidefinite. This in turn is true if $Z^{-2} - I$ is positive semidefinite. But

$$Z^{-2} - I = [I + k(X'X)^{-1}]^2 - I$$
$$= 2k(X'X)^{-1} + k(X'X)^{-2}$$

which is indeed positive semidefinite.

The above result shows that ridge regression is a way of "shrinking" the OLS estimator. This imparts a bias but reduces the variance. Another way of doing this is to simply multiply the OLS estimate by a fraction less than one.

38. PROPOSITION: Define $\beta^* = c\hat{\beta}$, c being a scalar. Then $E(\beta^* - \beta)'(\beta^* - \beta) \leq E(\hat{\beta} - \beta)'(\hat{\beta} - \beta)$ for c in the range

$$\frac{\beta'\beta - \sigma^2 \text{ trace}(X'X)^{-1}}{\beta'\beta + \sigma^2 \text{ trace}(X'X)^{-1}} \leq c \leq 1$$

Further, $E(\beta^* - \beta)'(\beta^* - \beta)$ is minimized at the midpoint of this range, namely at

$$c = \frac{\beta'\beta}{\beta'\beta + \sigma^2 \text{ trace}(X'X)^{-1}}$$

Proof: It is clear that

$$E(\hat{\beta} - \beta)'(\hat{\beta} - \beta) = \sigma^2 \text{ trace}(X'X)^{-1}$$

and

$$E(\beta^* - \beta)'(\beta^* - \beta) = (c - 1)^2\beta'\beta + c^2\sigma^2 \text{ trace}(X'X)^{-1}$$

Therefore, $E(\beta^* - \beta)'(\beta^* - \beta) \leq E(\hat{\beta} - \beta)'(\hat{\beta} - \beta)$ if and only if

$$(c - 1)^2\beta'\beta + c^2\sigma^2 \text{ trace}(X'X)^{-1} \leq \sigma^2 \text{ trace}(X'X)^{-1}$$

that is, if

$$[\beta'\beta + \sigma^2 \text{ trace}(X'X)^{-1}]c^2 - 2\beta'\beta c + [\beta'\beta - \sigma^2 \text{ trace}(X'X)^{-1}] \leq 0$$

The two roots of this quadratic are 1 and

$$\frac{\beta'\beta - \sigma^2 \text{ trace}(X'X)^{-1}}{\beta'\beta + \sigma^2 \text{ trace}(X'X)^{-1}}$$

Since the term on c^2 is positive, it is clear that negative values of the quadratic occur for values of c between these points. This proves the first part of the proposition.

To prove the second part, we minimize $E(\beta^* - \beta)'(\beta^* - \beta)$ with respect to c. This gives the first-order condition

$$2(c - 1)\beta'\beta + 2c\sigma^2 \text{ trace}(X'X)^{-1} = 0$$

which implies that the minimum occurs at

$$c = \frac{\beta'\beta}{\beta'\beta + \sigma^2 \text{ trace}(X'X)^{-1}}$$

39. COMMENT: Again we have a problem since the optimum value of c, and indeed the whole "beneficial" range of c, depends on unknown parameters. However, there are in the literature some results along these lines, where properties of such estimators (with estimates substituted for unknown parameters) have been obtained. See Bock (1975).

2.4 NONNORMAL DISTURBANCES

In this section we will suppose that all of the full ideal conditions hold except that the disturbances are not normally distributed. In particular, we still suppose that the disturbances ε_t are independently and identically distributed (iid) with zero mean and *finite* variance σ^2; however, we no longer suppose that their distribution is normal.

1. PROPOSITION: All the results of Section 1.3 are valid in this case. In particular, $\hat{\beta}$ is unbiased, consistent, BLUE, and has covariance matrix $\sigma^2(X'X)^{-1}$; s^2 is unbiased and consistent.

Proof: See Section 1.3.

2. PROPOSITION: The results of Section 1.4 are not valid (in general) when ε is not normally distributed. In particular, $\hat{\beta}$ does not have a normal distribution, $(T - K)s^2/\sigma^2$ does not have a chi-squared distribution, and neither $\hat{\beta}$ nor s^2 are efficient or asymptotically efficient.

Proof: The proofs of Section 1.4 depend on ε being normal.

3. PROPOSITON: The tests of hypotheses developed in Section 1.5 are not valid (in general) when ε is not normally distributed.

Proof: See Section 1.5.

Indeed, this fact is the crux of the problem. As Proposition 1 indicates, it is still possible to estimate β in a reasonable fashion - the OLS estimator $\hat{\beta}$ is unbiased, consistent, and indeed BLUE. However, one cannot make probability statements about these estimates since the distributions of $\hat{\beta}$ and of s^2 are not known. Of course, if the particular type of nonnormal distribution of ε were known, it might be possible to work out the distribution of $\hat{\beta}$ and of s^2, and hence to make the usual type of probability statements about the estimates. Whether or not this could actually be done would naturally depend on the tractability of the particular distribution [1].

In any case, the following results show that the usual test procedures are ***asymptotically*** justified (in a particular sense) whether or not the disturbances are normal.

4. LEMMA: If the elements of ε are iid with zero mean and finite variance σ^2, if the elements of X are uniformly bounded (i.e., there exists an R such that $|X_{ti}| \leq R$ for all t and i), and if

$$\underset{\sim}{Q} = \lim_{T\to\infty} \frac{X'X}{T}$$

is finite and nonsingular, then

$$\frac{1}{\sqrt{T}} X'\varepsilon$$

converges in distribution to $N(0,\sigma^2\underset{\sim}{Q})$. In somewhat simpler notation,

$$\frac{1}{\sqrt{T}} X'\varepsilon \to N(0,\sigma^2\underset{\sim}{Q})$$

Proof: We will consider first the case in which X is a vector, that is, there is only one regressor. We are therefore interested in the convergence of the univariate variable

$$Z_T = \frac{1}{\sqrt{T}} \sum_{t=1}^{T} X_t \varepsilon_t$$

Let F be the distribution function of ε_t (for any t) and G_t be the distribution function of $X_t \varepsilon_t$. Define S_T^2 by

$$S_T^2 = \sum_{t=1}^{T} \mathrm{Var}(X_t \varepsilon_t) = \sigma^2 \sum_{t=1}^{T} X_t^2$$

Note that in this case

$$\underset{\sim}{Q} = \lim_{T\to\infty} \frac{1}{T} \sum_{t=1}^{T} X_t^2$$

Then according to the Lindberg-Feller theorem (Theorem 2 of Section A.3), a necessary and sufficient condition for Z_t to converge to $N(0,\sigma^2 \underset{\sim}{Q})$ is that, for every $\varepsilon > 0$

$$\lim_{T\to\infty} \frac{1}{S_T^2} \sum_{t=1}^{T} \int_{|\omega|>\varepsilon S_T} \omega^2 dG_t(\omega) = 0$$

Now, using the fact that $G_t(\omega) = F(\omega/|X_t|)$, this requirement can be written as

$$\lim_{T\to\infty} \frac{T}{S_T^2} \sum_{t=1}^{T} \frac{X_t^2}{T} \int_{|\omega/X_t|>\varepsilon S_T/|X_t|} (\omega/X_t)^2 dF(\omega/|X_t|) = 0$$

Also, since

$$\lim_{T\to\infty} \frac{T}{S_T^2} = \frac{1}{\sigma^2 \underset{\sim}{Q}}$$

is finite and nonzero, we need only show that

$$\lim_{T\to\infty} \frac{1}{T} \sum_{t=1}^{T} X_t^2 \delta_{t,T} = 0$$

where

$$\delta_{t,T} = \int_{|\omega/X_t|>\varepsilon S_T/|X_t|} (\omega/X_t)^2 dF(\omega/|X_t|)$$

Now, it should be clear that

$$\lim_{T\to\infty} \delta_{t,T} = 0$$

for all t (and any fixed ε), since $|X_t|$ is bounded while S_T approaches infinity with T. Finally, since

$$\lim_{T\to\infty} \frac{1}{T} \sum_{t=1}^{T} X_t^2 = Q$$

is finite, and since

$$\lim_{T\to\infty} \delta_{t,T} = 0$$

for all t, it is clear that

$$\lim_{T\to\infty} \frac{1}{T} \sum_{t=1}^{T} X_t^2 \delta_{t,T} = 0$$

Hence the result follows, for the case in which X is composed of a single regressor.

We will now prove the lemma for the general case of K regressors, using Corollary 4 of Section A.3. To do so, pick an arbitrary K vector λ, and observe that

$$Z_t = \frac{\lambda' X' \varepsilon}{\sqrt{T}} = \frac{1}{\sqrt{T}} \sum_{t=1}^{T} \sum_{i=1}^{K} \lambda_i X_{ti} \varepsilon_t$$

But this is formally identical to the special case just considered, except that instead of X_t we now have in its place $\sum_{i=1}^{K} \lambda_i X_{ti}$. It therefore follows that the limiting distribution of Z_T is normal with zero mean and with variance

$$\lim_{T\to\infty} \frac{\sigma^2}{T} \sum_{t=1}^{T} \left(\sum_{i=1}^{K} \lambda_i X_{ti} \right)^2 = \lim_{T\to\infty} \frac{\sigma^2}{T} \sum_{t=1}^{T} \sum_{i=1}^{K} \sum_{j=1}^{K} \lambda_i \lambda_j X_{ti} X_{tj}$$

$$= \sigma^2 \sum_{i=1}^{K} \sum_{j=1}^{K} \lambda_i \lambda_j \lim_{T\to\infty} \frac{1}{T} \sum_{t=1}^{T} X_{ti} X_{tj}$$

$$= \sigma^2 \lambda' \underset{\sim}{Q} \lambda$$

where

$$\underset{\sim}{Q} = \lim_{T\to\infty} \frac{X'X}{T}$$

Finally, according to Corollary 4 of Section A.3, if

$$\frac{1}{\sqrt{T}} \lambda' X' \varepsilon \to N(0, \sigma^2 \lambda' \underset{\sim}{Q} \lambda)$$

for every λ,

$$\frac{1}{\sqrt{T}} X' \varepsilon \to N(0, \sigma^2 \underset{\sim}{Q})$$

The above lemma can be proved much more easily if we make the additional assumption that ε_t has a finite third absolute moment, say ρ^3. Then write

$$Z_T = \frac{\lambda' X' \varepsilon}{\sqrt{T}} = \frac{1}{\sqrt{T}} \sum_{t=1}^{T} w_t$$

where

$$w_t = \sum_{i=1}^{K} \lambda_i X_{ti} \varepsilon_t$$

But w_t clearly has zero mean, and its third absolute moment is bounded since ε_t has a finite third absolute moment and the X_{ti} are bounded. Also (in the notation of Theorem 7 of Section A.3),

$$A_t = E(w_t^2) = \sigma^2 \left(\sum_{i=1}^{K} \lambda_i X_{ti} \right)^2 = \sigma^2 \sum_{i=1}^{K} \sum_{j=1}^{K} \lambda_i \lambda_j X_{ti} X_{tj}$$

and

$$\lim_{H\to\infty} \frac{1}{H} \sum_{h=1}^{H} A_{t+h} = \sigma^2 \lambda' \underset{\sim}{Q} \lambda$$

for all t. Therefore the conditions of Theorem 7 of Section A.3 are satisfied (with m = 0), so that it follows that

$$\frac{1}{\sqrt{T}} \lambda' X' \varepsilon \to N(0, \sigma^2 \lambda' \underset{\sim}{Q} \lambda)$$

As before, this implies that

$$\frac{1}{\sqrt{T}} X' \varepsilon \to N(0, \sigma^2 \underset{\sim}{Q})$$

5. THEOREM: Under the conditions of Lemma 4,

$$\sqrt{T}(\hat{\beta} - \beta) \to N(0, \sigma^2 \underset{\sim}{Q}^{-1})$$

Proof: $\sqrt{T}(\hat{\beta} - \beta) = (X'X/T)^{-1}(1/\sqrt{T})X'\varepsilon$. Since $(X'X/T)^{-1} \to \underset{\sim}{Q}^{-1}$ and $(1/\sqrt{T})X'\varepsilon \to N(0, \sigma^2 \underset{\sim}{Q})$, it follows that

$$\sqrt{T}(\hat{\beta} - \beta) \to N(0, \sigma^2 \underset{\sim}{Q}^{-1} \underset{\sim}{Q} \underset{\sim}{Q}^{-1})$$

The above result shows that $\hat{\beta}$ has the same asymptotic distribution whether or not the disturbances are normal [see Note 2], as long as they are iid with zero mean and finite variance. The following results show that this implies that the usual t tests (Theorem 2 of Section 1.5) are asymptotically valid; the t distribution is of course replaced by the N(0,1) distribution, however.

6. PROPOSITION: Suppose that the elements of ε are iid with zero mean and finite variance, and that $\underset{\sim}{Q}$ is finite and nonsingular. Let R be a known $1 \times K$ vector, r a known scalar, and $\hat{\beta}$ the OLS estimator. Then under the null hypothesis that $R\beta = r$, the test statistic

$$\frac{R\hat{\beta} - r}{\sqrt{s^2 R(X'X)^{-1} R'}} \to N(0,1)$$

Proof: Under the null hypothesis,

$$R\hat{\beta} - r = R(\hat{\beta} - \beta)$$

so the test statistic is

$$\frac{\sqrt{T}\, R(\hat{\beta} - \beta)}{\sqrt{s^2 R(X'X/T)^{-1} R'}}$$

Now, since $\sqrt{T}(\hat{\beta} - \beta) \to N(0, \sigma^2 Q^{-1})$, it is clear that

$$\sqrt{T}\, R(\hat{\beta} - \beta) \to N(0, \sigma^2 R Q^{-1} R')$$

Also, the denominator of the test statistic has probability limit

$$\sqrt{\sigma^2 R Q^{-1} R'}$$

since $s^2 \to \sigma^2$ and $(X'X/T)^{-1} \to Q^{-1}$. But this is a scalar which is in fact the standard deviation of asymptotic distribution of the numerator random variable $\sqrt{T} R(\hat{\beta} - \beta)$. Hence the test statistic does indeed converge asymptotically to $N(0,1)$.

As should be clear, the test is only asymptotically valid. In any finite sample the normal distribution is only an approximation to the distribution of the test statistic. The t_{T-K} distribution can also be used; this is justified (asymptotically) since the t distribution converges to the $N(0,1)$ distribution as the number of degrees of freedom increases to infinity. Whether the t_{T-K} distribution or the $N(0,1)$ distribution is a better approximation (in *finite* samples) to the distribution of the above test statistic is an open question.

Note also that the above proof needed to be concerned only with the consistency of s^2 and not with any other feature of its asymptotic distribution. This is fortunate since the asymptotic distribution of s^2 does depend on the distribution of the disturbances, as the following proposition makes clear.

7. PROPOSITION: Let the elements of ε be iid with zero mean, finite variance σ^2, and finite fourth moment μ_4; assume also that Q is finite and nonsingular and X is uniformly bounded. Then

$$\sqrt{T}(s^2 - \sigma^2) \to N(0, \mu_4 - \sigma^4)$$

Proof: SSE = $\varepsilon' M \varepsilon$, where $M = I - X(X'X)^{-1}X'$, can be written as

$$SSE = \sum_{t=1}^{T-K} W_t^2$$

where $W = P'\varepsilon$ and P is the orthogonal matrix such that

$$P'MP = \begin{bmatrix} I & 0 \\ 0 & 0 \end{bmatrix} \qquad \text{[see Note 3]}$$

Since P is orthogonal, the W_t are iid with mean zero and variance σ^2, just like the ε_t. Thus

$$\frac{SSE}{\sigma^2} = \sum_{t=1}^{T-K} \left(\frac{W_t}{\sigma}\right)^2 = \sum_{t=1}^{T-K} V_t^2$$

where the V_t are iid with mean zero and variance one.

Clearly, then, the V_t^2 are iid with mean one and variance C^2 (to be evaluated later). Therefore,

$$\frac{1}{\sqrt{T - K}} \sum_{t=1}^{T-K} \frac{V_t^2 - 1}{C} \to N(0,1)$$

by the Lindberg-Levy central limit theorem (Theorem 1 of Section A.3). But this is just the statement that

$$\frac{1}{\sqrt{T - K}} \left[\frac{SSE}{\sigma^2} - (T - K)\right] \to N(0,C^2)$$

or that

$$\sqrt{T - K}(s^2 - \sigma^2) \to N(0,\sigma^4 C^2)$$

Clearly $\sqrt{T - K}$ and $\sqrt{T}$ are asymptotically equivalent. Furthermore,

$$C^2 = \mathrm{Var}(V_t^2) = E(V_t^4) - 1$$

But $E(V_t^4) = E(W_t/\sigma)^4 = E(\varepsilon_t/\sigma)^4 = \mu_4/\sigma^4$. Hence we have that

$$\sqrt{T}(s^2 - \sigma^2) \to N\left[0, \left(\frac{\mu_4}{\sigma^4} - 1\right)\sigma^4\right]$$

which is the result claimed.

It is worth noting that if the ε are normally distributed, then $\mu_4 - \sigma^4 = 2\sigma^4$; hence the above result is consistent with Proposition 14 of Section 1.4. Note also that the existence of the fourth moment of the disturbances is required.

Proposition 6 above showed that the usual t tests are asymptotically justified even if the disturbances are not normal, as long as they are iid. The following result shows that this is also true of the usual F tests.

8. PROPOSITION: Let the ε_t be iid with zero mean and finite variance σ^2. Let $\underset{\sim}{Q}$ be finite and nonsingular. Let R be a known m × 1 vector. Then under the null hypothesis that $R\beta = r$, the test statistic

$$\frac{(r - R\hat{\beta})'[R(X'X)^{-1}R']^{-1}(r - R\hat{\beta})/m}{SSE/(T - K)} \to \frac{\chi^2_m}{m}$$

Proof: The denominator of the test statistic is $SSE/(T - K) = s^2$, which is a consistent estimator of σ^2. Hence the asymptotic distribution of the statistic is that of the numerator divided by σ^2.

Under the null hypothesis, $r - R\hat{\beta} = -R(X'X)^{-1}X'\varepsilon$, so that the numerator of the test statistic can be expressed as $\varepsilon'Q\varepsilon/m$, where

$$Q = X(X'X)^{-1}R'[R(X'X)^{-1}R']^{-1}R(X'X)^{-1}X'$$

is symmetric and idempotent and has rank m [see Note 4]. But then we can write

$$\frac{\varepsilon'Q\varepsilon}{m\sigma^2} = \frac{\varepsilon'PP'QPP'\varepsilon}{m\sigma^2} = \frac{1}{m}V'\begin{bmatrix} I & 0 \\ 0 & 0 \end{bmatrix}V = \frac{1}{m}\sum_{i=1}^{m} V_i^2$$

where P is the orthogonal matrix such that $P'QP = \begin{bmatrix} I_m & 0 \\ 0 & 0 \end{bmatrix}$ and where $V = P'\varepsilon/\sigma$. Note that the V_i are iid with mean zero and variance one.

What has just been shown is that the test statistic has the asymptotic distribution of $1/m \sum_{i=1}^{m} V_i^2$. This in turn will be distributed as χ_m^2/m if the V_i are asymptotically normal.

Since $V = P'\varepsilon/\sigma$,

$$V_i = \sum_{t=1}^{T} \frac{P_{ti}\varepsilon_t}{\sigma} \qquad (i = 1, 2, \cdots, m)$$

Now the terms in this summation are independent variables with mean zero and variance $\sigma_t^2 = P_{ti}^2$. Also, since the ε_t are iid, it follows (as in the proof of Lemma 4) that the Lindberg-Feller central limit theorem is applicable. That is,

$$\sum_{t=1}^{T} \frac{P_{ti}\varepsilon_t/\sigma}{S_T} \to N(0,1)$$

where $S_T = \sqrt{\sum_{t=1}^{T} \sigma_t^2} = \sqrt{\sum_{t=1}^{T} P_{ti}^2} = 1$ since P is orthogonal. Therefore

$$\frac{1}{m} \sum_{i=1}^{m} V_i^2 \to \frac{\chi_m^2}{m}$$

In finite samples, of course, the χ_m^2/m distribution is only an approximation to the actual distribution of the test statistic. The $F_{m,T-K}$ distribution can also be used, since it converges to the χ_m^2/m distribution as T increases to infinity. Hence in this sense the usual F tests are asymptotically justified even if the disturbances are not normal, as long as they are iid with zero mean and finite variance.

2.5 GENERALIZED LEAST SQUARES

This section will assume that the full ideal conditions hold except that the covariance matrix of the disturbances is $\sigma^2\Omega$, where Ω is not the identity matrix. In particular, Ω may be nondiagonal and/or have unequal diagonal elements.

Properties of the Least Squares Estimators

1. PROPOSITION: The OLS estimator $\hat{\beta}$ is unbiased. Furthermore, if $\lim_{T\to\infty}(X'\Omega X/T)$ is finite, $\hat{\beta}$ is consistent.

Proof: $E(\hat{\beta}) = \beta + E(X'X)^{-1}X'\varepsilon = \beta$, which proves unbiasedness. Also

$$\text{plim}\ \hat{\beta} = \beta + \lim_{T\to\infty} \frac{X'X}{T}\ \text{plim}\ \frac{X'\varepsilon}{T}$$

But $X'\varepsilon/T$ has zero mean and covariance matrix

$$\sigma^2 \frac{X'\Omega X}{T^2}$$

If $\lim_{T\to\infty}(X'\Omega X/T)$ is finite, then $\lim_{T\to\infty}(X'\Omega X/T^2) = 0$. Hence $X'\varepsilon/T$ has zero mean and its covariance matrix vanishes asymptotically, which implies $\text{plim}(X'\varepsilon/T) = 0$, and therefore $\text{plim}\ \hat{\beta} = \beta$.

2. PROPOSITION: The covariance matrix of $\hat{\beta}$ is

$$\sigma^2(X'X)^{-1}X'\Omega X(X'X)^{-1}$$

Proof:

$$\begin{aligned} E(\hat{\beta} - \beta)(\hat{\beta} - \beta)' &= E(X'X)^{-1}X'\varepsilon\varepsilon'X(X'X)^{-1} \\ &= \sigma^2(X'X)^{-1}X'\Omega X(X'X)^{-1} \end{aligned}$$

3. COROLLARY: The asymptotic covariance matrix of $\hat{\beta}$ is

$$\frac{\sigma^2}{T} \lim_{T\to\infty}\left(\frac{X'X}{T}\right)^{-1} \lim_{T\to\infty} \frac{X'\Omega X}{T} \lim_{T\to\infty}\left(\frac{X'X}{T}\right)^{-1}$$

Note that the covariance matrix of $\hat{\beta}$ is no longer equal to $\sigma^2(X'X)^{-1}$. It may be *either* "larger" or "smaller," in the sense that $(X'X)^{-1}X'\Omega X(X'X)^{-1} - (X'X)^{-1}$ can be either positive semidefinite, negative semidefinite, or neither.

4. PROPOSITION: $s^2 = SSE/(T - K)$ is (in general) a biased and inconsistent estimator of σ^2.

Proof:

$$\begin{aligned} E(SSE) &= E(\varepsilon' M\varepsilon) \qquad [\text{where } M = I - X(X'X)^{-1}X'] \\ &= \text{trace } E(M\varepsilon\varepsilon') \\ &= \sigma^2 \text{trace } M\Omega \\ &\neq \sigma^2(T - K) \qquad \text{(in general)} \end{aligned}$$

Also, $\text{plim } s^2 = \sigma^2 \lim_{T\to\infty}(1/T)\text{trace } M\Omega \neq \sigma^2$.

The Generalized Least Squares (GLS) Estimators

5. LEMMA: There exists a nonsingular matrix V such that $V'V = \Omega^{-1}$.

Proof: Since Ω is a covariance matrix and is nonsingular, it is positive definite. This implies that Ω^{-1} is also positive definite, and hence a nonsingular V exists such that $V'V = \Omega^{-1}$.

6. PROPOSITION: Suppose that the regression equation $y = X\beta + \varepsilon$ satisfies the full ideal conditions except that Ω (the covariance matrix of the disturbances) is not the identity matrix. Suppose that

$$\lim_{T\to\infty} \frac{X'\Omega^{-1}X}{T}$$

is finite and nonsingular. Let V be a matrix such that $V'V = \Omega^{-1}$. Then the transformed equation

$$Vy = VX\beta + V\varepsilon$$

satisfies the full ideal conditions.

Proof: Since V is nonsingular and nonstochastic, VX is nonstochastic and of full rank if X is. Also,

$$\lim_{T\to\infty} \frac{(VX)'(VX)}{T} = \lim_{T\to\infty} \frac{X'\Omega^{-1}X}{T}$$

is finite and nonsingular by assumption. Therefore the transformed regressor matrix satisfies the required conditions, and we need consider only the transformed disturbance $V\varepsilon$.

Clearly $E(V\epsilon) = 0$. Also $E(V\epsilon)(V\epsilon)' = \sigma^2 V\Omega V'$. But $V'(V\Omega V') = \Omega^{-1}\Omega V' = V'$; since V' is nonsingular, this implies that $V\Omega V' = I$. Therefore $E(V\epsilon)(V\epsilon)' = \sigma^2 I$. Finally, the normality of $V\epsilon$ follows immediately from the normality of ϵ.

7. THEOREM (Aitken): The BLUE of β is

$$\tilde{\beta} = (X'\Omega^{-1}X)^{-1}X'\Omega^{-1}y$$

Proof: Since the transformed equation satisfies the full ideal conditions, the BLUE of β is just

$$\begin{aligned}\tilde{\beta} &= [(VX)'(VX)]^{-1}(VX)'(Vy)\\ &= (X'\Omega^{-1}X)^{-1}X'\Omega^{-1}y\end{aligned}$$

Indeed, since $\tilde{\beta}$ is the OLS estimator of β in the transformed equation, and since the transformed equation satisfies the ideal conditions, $\tilde{\beta}$ has all the usual desirable properties--it is unbiased, BLUE, efficient, consistent, and asymptotically efficient.

8. PROPOSITION: The covariance matrix of the GLS estimator $\tilde{\beta}$ is $\sigma^2(X'\Omega^{-1}X)^{-1}$.

Proof: Viewing $\tilde{\beta}$ as the OLS estimator in the transformed equation, it clearly has covariance matrix

$$\sigma^2[(VX)'(VX)]^{-1} = \sigma^2(X'\Omega^{-1}X)^{-1}$$

9. COROLLARY: The asymptotic covariance matrix of $\tilde{\beta}$ is

$$\frac{\sigma^2}{T}\lim_{T\to\infty}\left(\frac{X'\Omega^{-1}X}{T}\right)^{-1}$$

10. PROPOSITION: An unbiased, consistent, efficient, and asymptotically efficient estimator of σ^2 is

$$\tilde{\sigma}^2 = \frac{\tilde{\epsilon}'\Omega^{-1}\tilde{\epsilon}}{T - K}$$

where $\tilde{\epsilon} = y - X\tilde{\beta}$.

Proof: Since the transformed equation satisfies the ideal conditions, the desired estimator of σ^2 is

$$\frac{1}{T - K}(Vy - VX\tilde{\beta})'(Vy - VX\tilde{\beta}) = \frac{1}{T - K}(y - X\tilde{\beta})'\Omega^{-1}(y - X\tilde{\beta})$$

Finally, it should be clear that all the usual tests of hypotheses (such as those of Section 1.5) can be applied directly to the transformed equation. Equivalently, these tests could be developed by noting that

$$\tilde{\beta} \sim N[\beta, \sigma^2(X'\Omega^{-1}X)^{-1}]$$

that

$$\frac{(T - K)\tilde{\sigma}^2}{\sigma^2} \sim \chi^2_{T-K}.$$

and that $\tilde{\beta}$ and $\tilde{\sigma}^2$ are independent.

Case of Unknown Ω

This section has so far treated Ω as a known matrix. In fact, Ω is generally unknown. In such cases it is sometimes possible to use an estimator $\hat{\Omega}$ in place of Ω. The word sometimes is used advisedly, since Ω contains $(1/2)T(T + 1)$ distinct elements, all of which cannot be estimated from a sample of T observations.

Of course, Ω is usually assumed to depend on some fixed, finite number of parameters, and when these can be estimated one can form an estimator $\hat{\Omega}$ of Ω.

11. COMMENT: If Ω depends on a finite number of parameters $\theta_1, \theta_2, \cdots, \theta_p$, and if $\hat{\Omega}$ depends (in the same manner) on consistent estimators $\hat{\theta}_1, \hat{\theta}_2, \cdots, \hat{\theta}_p$ of these parameters, then $\hat{\Omega}$ is sometimes called a consistent estimator of Ω.

Alternatively, $\hat{\Omega}$ is sometimes called a consistent estimator of Ω if $\text{plim}\, \hat{\Omega}_{ij} = \Omega_{ij}$ for all i and j [see Note 1].

12. DEFINITION: Let $\hat{\Omega}$ be an estimator of Ω. Then the *feasible generalized least squares* (feasible GLS) estimator of β is

$$\bar{\beta} = (X'\hat{\Omega}^{-1}X)^{-1}X'\hat{\Omega}^{-1}y$$

It is sometimes alleged that, if $\hat{\Omega}$ is a consistent estimator of Ω (in the sense of Comment 11), the feasible GLS estimator has the same asymptotic distribution as the GLS estimator itself. [This supposed result is sometimes mistakenly attributed to Zellner (1962).] This is not true, as is illustrated by the following example.

13. EXAMPLE: Let

$$X = \begin{bmatrix} 1 \\ 1 \\ 1 \\ \cdot \\ \cdot \\ \cdot \\ 1 \end{bmatrix} \quad \text{and} \quad \Omega^{-1} = \begin{bmatrix} 1 & & & & 0 \\ & \lambda & & & \\ & & \lambda^2 & & \\ & & & \ddots & \\ 0 & & & & \lambda^{T-1} \end{bmatrix}$$

Let $\hat{\lambda} = \lambda + 1/T$, and let $\lambda = 1$ in this example. Note that plim $\hat{\lambda} = \lambda$, and that plim $\hat{\lambda}^i = \lambda^i$ for all i, so that $\hat{\Omega}$ is a consistent estimator of Ω (in the sense of Comment 11). Then

$$\frac{X'\Omega^{-1}X}{T} = 1$$

and

$$\frac{X'\Omega^{-1}\varepsilon}{\sqrt{T}} = \frac{1}{\sqrt{T}} \sum_{t=1}^{T} \varepsilon_t \to N(0,\sigma^2)$$

Therefore

$$\sqrt{T}(\tilde{\beta} - \beta) = \left(\frac{X'\Omega^{-1}X}{T}\right)^{-1} \frac{X'\Omega^{-1}\varepsilon}{\sqrt{T}} \to N(0,\sigma^2)$$

On the other hand,

$$\frac{X'\hat{\Omega}^{-1}X}{T} = \frac{1}{T} \sum_{t=1}^{T} \left(1 + \frac{1}{T}\right)^{t-1}$$

$$= \frac{1}{T} \frac{1 - (1 + 1/T)^T}{1 - (1 + 1/T)} \to e - 1$$

since $\lim_{T\to\infty}(1 + 1/T)^T = e$, the base of natural logarithms. Also,

$$\frac{X'\hat{\Omega}^{-1}\varepsilon}{\sqrt{T}} = \frac{1}{\sqrt{T}} \sum_{t=1}^{T} \varepsilon_t \left(1 + \frac{1}{T}\right)^{t-1}$$

is normally distributed with zero mean and variance

$$\begin{aligned}
\frac{\sigma^2}{T} \sum_{t=1}^{T} \left(1 + \frac{1}{T}\right)^{2(t-1)} &= \frac{\sigma^2}{T} \frac{1 - (1 + 1/T)^{2T}}{1 - (1 + 1/T)^2} \\
&= \frac{\sigma^2\{1 - [(1 + 1/T)^T]^2\}}{T[1 - 1 - 2/T - 1/T^2]} \\
&\to \frac{\sigma^2(e^2 - 1)}{2}
\end{aligned}$$

Therefore,

$$\sqrt{T}(\bar{\beta} - \beta) = \left(\frac{X'\hat{\Omega}^{-1}X}{T}\right)^{-1} \frac{X'\hat{\Omega}^{-1}\varepsilon}{\sqrt{T}} \to N\left[0, \sigma^2 \frac{e^2 - 1}{2(e - 1)^2}\right]$$

Hence $\tilde{\beta}$ and $\bar{\beta}$ do not have the same asymptotic distribution. In fact, $(e^2 - 1)/[2(e - 1)^2]$ is approximately 1.25, so that using the consistent "estimator" $\hat{\lambda}$ instead of the true value of λ has increased the asymptotic variance by about 25%.

Indeed, there is no general rule as to when the feasible GLS estimator and the GLS estimator have the same asymptotic distribution--this is something to be checked in every case. The following results provide a method of checking this.

14. PROPOSITION: A *sufficient* condition for the feasible GLS estimator to be consistent is that

$$\operatorname{plim} \frac{X'\hat{\Omega}^{-1}X}{T}$$

be finite and nonsingular, and that

$$\operatorname{plim} \frac{X'\hat{\Omega}^{-1}\varepsilon}{T} = 0$$

Proof: $\bar{\bar{\beta}} = \beta + (X'\hat{\Omega}^{-1}X)^{-1}X'\hat{\Omega}^{-1}\varepsilon$. Hence

$$\operatorname{plim} \bar{\bar{\beta}} = \beta + \operatorname{plim}\left(\frac{X'\hat{\Omega}^{-1}X}{T}\right)^{-1} \operatorname{plim} \frac{X'\hat{\Omega}^{-1}\varepsilon}{T}$$

and the result follows.

15. THEOREM: A *sufficient* condition for the GLS estimator and the feasible GLS estimator to have the same asymptotic distributions is that

$$\operatorname{plim} \frac{X'(\hat{\Omega}^{-1} - \Omega^{-1})X}{T} = 0$$

and

$$\operatorname{plim} \frac{X'(\hat{\Omega}^{-1} - \Omega^{-1})\varepsilon}{\sqrt{T}} = 0$$

Proof: We have

$$\sqrt{T}(\tilde{\beta} - \beta) = \left(\frac{X'\Omega^{-1}X}{T}\right)^{-1} \frac{X'\Omega^{-1}\varepsilon}{\sqrt{T}}$$

and

$$\sqrt{T}(\bar{\bar{\beta}} - \beta) = \left(\frac{X'\hat{\Omega}^{-1}X}{T}\right) \frac{X'\hat{\Omega}^{-1}\varepsilon}{\sqrt{T}}$$

Therefore we have $\operatorname{plim}\sqrt{T}(\tilde{\beta} - \bar{\bar{\beta}}) = 0$ if

$$\operatorname{plim} \frac{X'\Omega^{-1}X}{T} = \operatorname{plim} \frac{X'\hat{\Omega}^{-1}X}{T}$$

and

$$\operatorname{plim} \frac{X'\Omega^{-1}\varepsilon}{\sqrt{T}} = \operatorname{plim} \frac{X'\hat{\Omega}^{-1}\varepsilon}{\sqrt{T}}$$

But these are just restatements of the conditions of the theorem.

The conditions of Proposition 14 and Theorem 15 seem to be met in practice when $\hat{\Omega}$ is a consistent estimate of Ω, at least in a number of usual cases (seemingly unrelated regressions, autoregressive schemes of finite order, etc.). However, as indicated earlier, this is something that needs to be checked.

The small-sample properties of the feasible GLS estimators are generally unknown. One exception is the following result, which is occasionally useful.

16. PROPOSITION: Let $\hat{\Omega}$ be an estimator of Ω, the covariance matrix of the disturbance ε. Suppose that $\hat{\Omega}$ is an even function of ε [see Note 2]. Then the distribution of $\bar{\beta}$ is symmetric around β; $\bar{\beta}$ is unbiased if its mean exists.

Proof: $\bar{\beta} = \beta + (X'\hat{\Omega}^{-1}X)^{-1}X'\hat{\Omega}^{-1}\varepsilon$. Since ε is symmetrically distributed, ε and $-\varepsilon$ have equal densities. Since ε and $-\varepsilon$ imply the same value of $\hat{\Omega}$, it follows that

$$\bar{\beta} - \beta = (X'\hat{\Omega}^{-1}X)^{-1}X'\hat{\Omega}^{-1}\varepsilon$$

and

$$\beta - \bar{\beta} = (X'\hat{\Omega}^{-1}X)^{-1}X'\hat{\Omega}^{-1}(-\varepsilon)$$

have equal densities. Therefore $\bar{\beta}$ is distributed symmetrically around β [see Note 3].

2.6 SETS OF REGRESSION EQUATIONS

Let us consider a set of G regression equations:

$$y_1 = X_1\beta_1 + \varepsilon_1$$

$$y_2 = X_2\beta_2 + \varepsilon_2$$

$$\cdots\cdots$$

$$y_G = X_G\beta_G + \varepsilon_G$$

where each y_i $(i = 1, 2, \cdots, G)$ is of dimension $T \times 1$, X_i is $T \times K_i$, β_i is $K_i \times 1$, and ε_i is $T \times 1$. Assume that each individual equation satisfies the full ideal condition.

1. NOTATION: The above set of equations can be written as one *combined equation*

$$y_* = X_*\beta_* + \varepsilon_*$$

where

$$y_* = \begin{pmatrix} y_1 \\ y_2 \\ \vdots \\ y_G \end{pmatrix} \quad \text{is of dimension } GT \times 1$$

$$X_* = \begin{pmatrix} X_1 & 0 & 0 & \cdots & 0 \\ 0 & X_2 & 0 & \cdots & 0 \\ 0 & 0 & X_3 & \cdots & 0 \\ \vdots & \vdots & \vdots & \ddots & \vdots \\ 0 & 0 & 0 & \cdots & X_G \end{pmatrix} \quad \text{is } GT \times \sum_{j=1}^{G} K_j$$

$$\beta_* = \begin{pmatrix} \beta_1 \\ \beta_2 \\ \vdots \\ \beta_G \end{pmatrix} \quad \text{is } \sum_{j=1}^{G} K_j \times 1$$

$$\varepsilon_* = \begin{pmatrix} \varepsilon_1 \\ \varepsilon_2 \\ \vdots \\ \varepsilon_G \end{pmatrix} \quad \text{is } GT \times 1$$

Note that this combined equation need *not* satisfy the full ideal conditions even though each equation does separately. The reason is that the elements of ε_i may have a different variance than the elements of ε_j, and also that some elements of ε_i may be correlated with elements of ε_j $(i \neq j)$.

2. PROPOSITION: Let $\hat{\beta}_i = (X_i'X_i)^{-1}X_i'y_i$ $(i = 1, 2, \cdots, G)$ and let $\hat{\beta}_* = (X_*'X_*)^{-1}X_*'y_*$. Then

$$\hat{\beta}_* = \begin{bmatrix} \hat{\beta}_1 \\ \hat{\beta}_2 \\ \vdots \\ \hat{\beta}_G \end{bmatrix}$$

That is, OLS applied to the combined equation is identical to OLS applied to each equation separately.

Proof: $\hat{\beta}_* = (X_*'X_*)^{-1}X_*'y$

$$= \begin{bmatrix} X_1'X_1 & 0 & \cdots & 0 \\ 0 & X_2'X_2 & \cdots & 0 \\ \vdots & \vdots & \ddots & \vdots \\ 0 & 0 & \cdots & X_G'X_G \end{bmatrix}^{-1} \begin{bmatrix} X_1'y_1 \\ X_2'y_2 \\ \vdots \\ X_G'y_G \end{bmatrix}$$

$$= \begin{bmatrix} (X_1'X_1)^{-1}X_1'y_1 \\ \vdots \\ (X_G'X_G)^{-1}X_G'y_G \end{bmatrix} = \begin{bmatrix} \hat{\beta}_1 \\ \vdots \\ \hat{\beta}_G \end{bmatrix}$$

3. COROLLARY: A sufficient (but not necessary) condition for $\hat{\beta}_*$ to b an unbiased and consistent estimator of β_* is that each equation separately satisfy the full ideal conditions.

We now consider a case in which ε_* does not satisfy the full ideal conditions.

4. ASSUMPTION: Assume that each equation separately satisfies the full ideal conditions. Assume in addition that

$$\lim_{T\to\infty} \frac{X_i'X_j}{T}$$

is finite and nonsingular $(i,j = 1, 2, \cdots, G)$ and that

$$E(\varepsilon_i\varepsilon_j') = \sigma_{ij}I_T \qquad (i,j = 1, 2, \cdots, G)$$

where σ_{ij} is a finite scalar.

5. PROPOSITION: Let

$$\Sigma = \begin{pmatrix} \sigma_{11} & \sigma_{12} & \cdots & \sigma_{1G} \\ \vdots & \vdots & & \vdots \\ \sigma_{G1} & \sigma_{G2} & \cdots & \sigma_{GG} \end{pmatrix}$$

and let

$$\Omega = \Sigma \otimes I_T = \begin{pmatrix} \sigma_{11}I_T & \sigma_{12}I_T & \cdots & \sigma_{1G}I_T \\ \vdots & \vdots & & \vdots \\ \sigma_{G1}I_T & \sigma_{G2}I_T & \cdots & \sigma_{GG}I_T \end{pmatrix}$$

(Clearly Ω is of dimension $GT \times GT$.) Then Ω is the covariance matrix of ε_*.

Proof:

$$E(\varepsilon_*\varepsilon_*') = E\begin{pmatrix} \varepsilon_1\varepsilon_1' & \varepsilon_1\varepsilon_2' & \cdots & \varepsilon_1\varepsilon_G' \\ \varepsilon_2\varepsilon_1' & \varepsilon_2\varepsilon_2' & \cdots & \varepsilon_2\varepsilon_G' \\ \vdots & \vdots & & \vdots \\ \varepsilon_G\varepsilon_1' & \varepsilon_G\varepsilon_2' & \cdots & \varepsilon_G\varepsilon_G' \end{pmatrix}$$

$$= \begin{pmatrix} \sigma_{11}I & \sigma_{12}I & \cdots & \sigma_{1G}I \\ \sigma_{21}I & \sigma_{22}I & \cdots & \sigma_{2G}I \\ \vdots & \vdots & & \vdots \\ \sigma_{G1}I & \sigma_{G2}I & \cdots & \sigma_{GG}I \end{pmatrix}$$

$$= \Omega$$

6. COROLLARY: An unbiased, BLUE, efficient, consistent and asymptotically efficient estimator of β_* is

$$\tilde{\beta}_* = (X_*'\Omega^{-1}X_*)^{-1}X_*'\Omega^{-1}y_*$$

its covariance matrix is

$$(X_*'\Omega^{-1}X_*)^{-1}$$

and its asymptotic covariance matrix is

$$\frac{1}{T}\lim_{T\to\infty}\left(\frac{X'\Omega^{-1}X}{T}\right)^{-1}$$

7. ALGEBRAIC FACTS:

(a) $$\Omega^{-1} = \Sigma^{-1} \otimes I_T = \begin{bmatrix} \sigma^{11}I_T & \sigma^{12}I_T & \cdots & \sigma^{1G}I_T \\ \vdots & \vdots & & \vdots \\ \sigma^{G1}I_T & \sigma^{G2}I_T & \cdots & \sigma^{GG}I_T \end{bmatrix}$$

where σ^{ij} is the i,j-th element of Σ^{-1}.

(b) $X_*'\Omega^{-1}X_*$

$$= \begin{bmatrix} X_1 & 0 & \cdots & 0 \\ 0 & X_2 & \cdots & 0 \\ \vdots & \vdots & \ddots & \vdots \\ 0 & 0 & \cdots & X_G \end{bmatrix}' \begin{bmatrix} \sigma^{11}I_T & \sigma^{12}I_T & \cdots & \sigma^{1G}I_T \\ \vdots & \vdots & & \vdots \\ \sigma^{G1}I_T & \sigma^{G2}I_T & \cdots & \sigma^{GG}I_T \end{bmatrix} \begin{bmatrix} X_1 & 0 & \cdots & 0 \\ 0 & X_2 & \cdots & 0 \\ \vdots & \vdots & \ddots & \vdots \\ 0 & 0 & \cdots & X_G \end{bmatrix}$$

$$= \begin{bmatrix} \sigma^{11}X_1'X_1 & \sigma^{12}X_1'X_2 & \cdots & \sigma^{1G}X_1'X_G \\ \vdots & \vdots & & \vdots \\ \sigma^{G1}X_G'X_1 & \sigma^{G2}X_G'X_2 & \cdots & \sigma^{GG}X_G'X_G \end{bmatrix}$$

(which is clearly square and of dimension $\sum_{j=1}^{G} K_j$).

(c) $$X_*'\Omega^{-1}y_* = \begin{bmatrix} \sum_{j=1}^{G}\sigma^{1j}X_1'y_j \\ \sum_{j=1}^{G}\sigma^{2j}X_2'y_j \\ \vdots \\ \sum_{j=1}^{G}\sigma^{Gj}X_G'y_j \end{bmatrix}$$

8. COROLLARY: The estimator $\tilde{\beta}_*$ of Corollary 6 can be written as

$$\tilde{\beta} = \begin{pmatrix} \sigma^{11}X_1'X_1 & \sigma^{12}X_1'X_2 & \cdots & \sigma^{1G}X_1'X_G \\ \vdots & \vdots & \ddots & \vdots \\ \sigma^{G1}X_G'X_1 & \sigma^{G2}X_G'X_2 & \cdots & \sigma^{GG}X_G'X_G \end{pmatrix}^{-1} \begin{pmatrix} \sum_{j=1}^{G}\sigma^{1j}X_1'y_j \\ \vdots \\ \sum_{j=1}^{G}\sigma^{Gj}X_G'y_j \end{pmatrix}$$

The generalized least squares estimator $\tilde{\beta}_*$ is of course generally more efficient than the OLS estimator $\hat{\beta}_*$. However, there are two cases in which the two estimators are *identical*. The following is one of these two cases.

9. PROPOSITION: If $\sigma_{jk} = 0$ for all $j \neq k$, then $\tilde{\beta}_* = \hat{\beta}_*$. (In other words, the OLS estimator is fully efficient in this case.)

Proof: If $\sigma_{jk} = 0$ for all $j \neq k$, then $\tilde{\beta}_*$ in Corollary 8 reduces to

$$\beta_* = \begin{pmatrix} \sigma^{11}X_1'X_1 & 0 & \cdots & 0 \\ 0 & \sigma^{22}X_2'X_2 & \cdots & 0 \\ \vdots & \vdots & \ddots & \vdots \\ 0 & 0 & \cdots & \sigma^{GG}X_G'y_G \end{pmatrix}^{-1} \begin{pmatrix} \sigma^{11}X_1'y_1 \\ \sigma^{22}X_2'y_2 \\ \vdots \\ \sigma^{GG}X_G'y_G \end{pmatrix}$$

$$= \begin{pmatrix} (X_1'X_1)^{-1}X_1'y_1 \\ (X_2'X_2)^{-1}X_2'y_2 \\ \vdots \\ (X_G'X_G)^{-1}X_G'y_G \end{pmatrix}$$

This result should be intuitively appealing. When the disturbances are not correlated across equations, there is no reason why the efficiency of estimation of any particular equation should be improved by the use of information from another equation.

Note, however, that even in this case the combined disturbance ε_* does not satisfy the full ideal conditions, since σ_{ii} need not equal σ_{jj}. Nevertheless OLS is efficient in this case.

Identical Regressors

In this subsection we will suppose that the regressor matrix in each equation is the same. That is, $X_i = X$ for all $i = 1, 2, \cdots, G$.

10. PROPOSITION: Consider the set of equations

$$\begin{aligned} y_1 &= X\beta_1 + \varepsilon_1 \\ y_2 &= X\beta_2 + \varepsilon_2 \\ &\cdots\cdots \\ y_G &= X\beta_G + \varepsilon_G \end{aligned}$$

where the assumptions made are as before (Assumption 4). Then OLS applied to each equation is fully efficient.

Proof: As before, the efficient estimator is

$$\tilde{\beta}_* = (X_*'\Omega^{-1}X_*)^{-1}X_*'\Omega^{-1}y$$

the notation being as previously defined. But in this case

$$X_*'\Omega^{-1}X_* = \begin{pmatrix} \sigma^{11}X'X & \sigma^{12}X'X & \cdots & \sigma^{1G}X'X \\ \vdots & \vdots & & \vdots \\ \sigma^{G1}X'X & \sigma^{G2}X'X & & \sigma^{GG}X'X \end{pmatrix} = \Sigma^{-1} \otimes X'X$$

so

$$(X_*'\Omega^{-1}X_*)^{-1} = \Sigma \otimes (X'X)^{-1}$$

$$= \begin{pmatrix} \sigma_{11}(X'X)^{-1} & \sigma_{12}(X'X)^{-1} & \cdots & \sigma_{1G}(X'X)^{-1} \\ \vdots & \vdots & \ddots & \vdots \\ \sigma_{G1}(X'X)^{-1} & \sigma_{G2}(X'X)^{-1} & \cdots & \sigma_{GG}(X'X)^{-1} \end{pmatrix}$$

Also

$$X_*'\Omega^{-1}y_* = \begin{pmatrix} \sum_{j=1}^{G}\sigma^{1j}X'y_j \\ \vdots \\ \sum_{j=1}^{G}\sigma^{Gj}X'y_j \end{pmatrix}$$

Then the p-th vector element of $\tilde{\beta}_*$ is by definition the p-th subvector of

$$(X_*'\Omega^{-1}X_*)^{-1}X_*'\Omega^{-1}y_*$$

which is just (by the definition of block matrix multiplication)

$$\begin{aligned}\tilde{\beta}_{*(p)} &= \sum_{k=1}^{G}\sigma_{pk}(X'X)^{-1}\sum_{j=1}^{G}\sigma^{kj}X'y_j\\ &= \sum_{j=1}^{G}\left[\sum_{k=1}^{G}\sigma_{pk}\sigma^{kj}\right](X'X)^{-1}X'y_j\end{aligned}$$

Now

$$\sum_{k=1}^{G}\sigma_{pk}\sigma^{kj} = \delta_{pj}, \quad \text{where} \quad \delta_{pj} = \begin{cases}1 \text{ if } p = j\\ 0 \text{ if } p \neq j\end{cases}$$

is just the p,j-th element of $\Sigma\Sigma^{-1} = I$. Hence,

$$\begin{aligned}\tilde{\beta}_{*(p)} &= \sum_{j=1}^{G}\delta_{pj}(X'X)^{-1}X'y_j\\ &= (X'X)^{-1}X'y_p\\ &= \hat{\beta}_p\end{aligned}$$

the OLS estimator in the p-th equation.

Hence in this case too the joint GLS procedure is unnecessary. This is true even though there may be correlation across equations ($\sigma_{kj} \neq 0$ for $k \neq j$).

11. NOTATION: Define

$Y = (y_1 y_2 \cdots y_G)$ which is of dimension $T \times G$

$\beta = (\beta_1 \beta_2 \cdots \beta_G)$ which is $K \times G$

$\varepsilon = (\varepsilon_1 \varepsilon_2 \cdots \varepsilon_G)$ which is $T \times G$

Then the set of equations

$$y_i = X\beta_i + \varepsilon_i \qquad (i = 1, 2, \cdots, G)$$

can be written as

$$Y = X\beta + \varepsilon$$

12. COMMENT: $\hat{\beta} = (X'X)^{-1}X'Y = (\hat{\beta}_1\hat{\beta}_2 \cdots \hat{\beta}_G)$ just amounts to OLS applied to each equation separately; in other words, the p-th column of $\hat{\beta}$ is $\hat{\beta}_p = (X'X)^{-1}X'y_p$. Also note that by Theorem 10, $\hat{\beta}$ is an efficient estimator of β.

An interesting fact, though of no particular importance in the present context, is the following.

13. THEOREM: Let $E = Y - X\hat{\beta}$ and $W = E'E$. Then the value of $\hat{\beta}$ which minimizes $|W|$ is $\hat{\beta} = (X'X)^{-1}XY$, the OLS estimator.

Proof: We will proceed by minimizing $\log|W|$. Now

$$\frac{\partial \log|W|}{\partial\beta_{ij}} = \sum_{r=1}^{G}\sum_{s=1}^{G}\frac{\partial \log|W|}{\partial w_{rs}}\frac{\partial w_{rs}}{\partial\beta_{ij}} \qquad (i = 1, 2, \cdots, K;\ j = 1, 2, \cdots, G)$$

where w_{rs} is the r,s-th element of W. Using the fact that $(\partial \log|W|)/\partial w_{rs} = w^{sr}$, the s,r-th element of W^{-1}, we have

$$\frac{\partial \log|W|}{\partial\beta_{ij}} = \sum_{r=1}^{G}\sum_{s=1}^{G} w^{sr}\frac{\partial w_{rs}}{\partial\beta_{ij}}$$

$$= \sum_{r=1}^{G}\sum_{s=1}^{G} w^{rs}\frac{\partial w_{rs}}{\partial\beta_{ij}} \qquad \text{(since } W^{-1} \text{ is symmetric)}$$

But $w_{rs} = \sum_{t=1}^{T}e_{tr}e_{ts}$ where e_{ts} is the t,s-th element of E; therefore,

$$\frac{\partial w_{rs}}{\partial\beta_{ij}} = \sum_{t=1}^{T}\left(e_{ts}\frac{\partial e_{tr}}{\partial\beta_{ij}} + e_{tr}\frac{\partial e_{ts}}{\partial\beta_{ij}}\right)$$

Since $e_{ts} = y_{ts} - \sum_{k=1}^{K}X_{tk}\beta_{ks}$, we have

$$\frac{\partial e_{ts}}{\partial\beta_{ij}} = \begin{cases} -X_{ti} & \text{if } j = s \\ 0 & \text{if } j \neq s \end{cases}$$

Therefore,

$$\frac{\partial w_{rs}}{\partial \beta_{ij}} = \begin{cases} 0 & \text{if } r \neq j \text{ and } s \neq j \\ -\sum_{t=1}^{T} e_{ts} X_{ti} & \text{if } r = j \text{ but } s \neq j \\ -\sum_{t=1}^{T} e_{tr} X_{ti} & \text{if } s = j \text{ but } r \neq j \\ -2\sum_{t=1}^{T} e_{ts} X_{ti} & \text{if } r = s = j \end{cases}$$

Hence

$$\sum_{r=1}^{G} \sum_{s=1}^{G} w^{rs} \frac{\partial w_{rs}}{\partial \beta_{ij}} = -2w^{jj} \sum_{t=1}^{T} e_{tj} X_{ti} - \sum_{r \neq j} w^{rj} \sum_{t=1}^{T} e_{tr} X_{ti} - \sum_{s \neq j} w^{js} \sum_{t=1}^{T} e_{ts} X_{ti}$$

where the first term corresponds to r = s = j, the second to s = j but r ≠ j, and the third to r = j but s ≠ j. Now in light of the symmetry of W^{-1} this is just

$$-2w^{jj} \sum_{t=1}^{T} e_{tj} X_{ti} - 2 \sum_{r \neq j} w^{rj} \sum_{t=1}^{T} e_{tr} X_{ti} = -2 \sum_{r=1}^{G} w^{rj} \sum_{t=1}^{T} e_{tr} X_{ti}$$

But $\sum_{t=1}^{T} e_{tr} X_{ti}$ is just the i,r-th element of X'E, say $(X'E)_{ir}$, and so

$$\frac{\partial \log |W|}{\partial \beta_{ij}} = -2 \sum_{r=1}^{G} (X'E)_{ir} w^{rj}.$$

Finally, this is just -2 times the i,j-th element of $X'EW^{-1}$; that is,

$$\frac{\partial \log |W|}{\partial \beta_{ij}} = -2(X'EW^{-1})_{ij}$$

Setting this equal to zero (to find the minimum) and postmultiplying by W gives

$$-2(X'Y - X'X\beta) = 0$$

the solution of which is indeed

$$\beta = (X'X)^{-1}X'Y = \hat{\beta}$$

Case of Unknown Covariance Matrix Σ

Let us return to the case where the regressor matrix may differ from equation to equation. Explicitly, we return to the case of

$$y_i = X_i\beta_i + \varepsilon_i \qquad (i = 1, 2, \cdots, G)$$

and the combined equation

$$y_* = X_*\beta_* + \varepsilon_*$$

these being defined as in Notation 1. Again let

$$\Omega = E(\varepsilon_*\varepsilon_*') = \Sigma \otimes I$$

In this subsection, however, we will suppose that Σ is unknown and must be replaced with an estimator.

14. PROPOSITION: Let $e_j = y_j - X_j(X_j'X_j)^{-1}X_j'y_j$ be the set of OLS residuals from the j-th equation. Then a consistent estimator of σ_{ij} is

$$s_{ij} = \frac{1}{T} e_i'e_j$$

Proof:

$$\frac{1}{T} e_i'e_j = \frac{1}{T} \varepsilon_i'[I - X_i(X_i'X_i)^{-1}X_i'][I - X_j(X_j'X_j)^{-1}X_j']\varepsilon_j$$

$$= \frac{\varepsilon_i'\varepsilon_j}{T} - \frac{\varepsilon_i'X_i}{T}\left(\frac{X_i'X_i}{T}\right)^{-1}\frac{X_i'\varepsilon_j}{T} - \frac{\varepsilon_i'X_j}{T}\left(\frac{X_j'X_j}{T}\right)^{-1}\frac{X_j'\varepsilon_j}{T}$$

$$+ \frac{\varepsilon_i'X_i}{T}\left(\frac{X_i'X_i}{T}\right)^{-1}\frac{X_i'X_j}{T}\left(\frac{X_j'X_j}{T}\right)^{-1}\frac{X_j'\varepsilon_j}{T}$$

Clearly $\text{plim}(\varepsilon_i'\varepsilon_j/T) = \sigma_{ij}$. Therefore what must be shown is that the last three terms have zero probability limits. But this should be clear since

$$\lim_{T\to\infty}\frac{X_i'X_i}{T}, \quad \lim_{T\to\infty}\frac{X_j'X_j}{T}, \quad \text{and} \quad \lim_{T\to\infty}\frac{X_i'X_j}{T}$$

are finite and nonsingular, while

$$\operatorname{plim} \frac{X_i'\varepsilon_i}{T}, \quad \operatorname{plim} \frac{X_j'\varepsilon_j}{T}, \quad \operatorname{plim} \frac{X_i'\varepsilon_j}{T}, \quad \text{and} \quad \operatorname{plim} \frac{X_j'\varepsilon_i}{T}$$

are all zero.

15. NOTATION: Let

$$S = \begin{pmatrix} s_{11} & s_{12} & \cdots & s_{1G} \\ \vdots & \vdots & & \vdots \\ s_{G1} & s_{G2} & \cdots & s_{GG} \end{pmatrix} \qquad (s_{ij} \text{ defined in Proposition 14})$$

and $\hat{\Omega} = S \otimes I_T$. Let

$$\bar{\beta}_* = (X_*'\hat{\Omega}^{-1}X_*)^{-1}X_*'\hat{\Omega}^{-1}y_*$$

and, as before,

$$\tilde{\beta}_* = (X_*'\Omega^{-1}X_*)^{-1}X_*'\Omega^{-1}y_*$$

16. LEMMA: $\operatorname{plim} \dfrac{X_*'(\hat{\Omega}^{-1}-\Omega^{-1})X_*}{T} = 0$

Proof: $$\frac{X_*'(\hat{\Omega}^{-1}-\Omega^{-1})X_*}{T} = \frac{X_*'[(S^{-1}-\Sigma^{-1}) \otimes I]X_*}{T}$$

Using Algebraic Fact 7(b), the i,j-th submatrix of this is just

$$(s^{ij} - \sigma^{ij})\frac{X_i'X_j}{T}$$

where s^{ij} and σ^{ij} are the i,j=th elements of S^{-1} and Σ^{-1}. But since $\operatorname{plim} s^{ij} = \sigma^{ij}$, we have

$$\operatorname{plim}(s^{ij} - \sigma^{ij}) \frac{X_i'X_j}{T} = \operatorname{plim}(s^{ij} - \sigma^{ij}) \cdot \lim \frac{X_i'X_j}{T} = 0$$

and the lemma follows.

17. COROLLARY: $\operatorname{plim} \dfrac{X_*'\hat{\Omega}^{-1}X_*}{T}$ is finite and nonsingular.

18. LEMMA: $\text{plim}\,\dfrac{X_*'\hat{\Omega}^{-1}\varepsilon_*}{T} = 0$

Proof: Using a slight modification of Algebraic Fact 7(c), the i-th subvector of the above vector is

$$\frac{1}{T}\sum_{j=1}^{G} s^{ij}X_i'\varepsilon_j$$

whose probability limit is

$$\sum_{j=1}^{G} \sigma^{ij}\text{plim}\,\frac{X_i'\varepsilon_j}{T} = 0$$

19. PROPOSITION: The estimator

$$\bar{\beta}_* = (X_*'\hat{\Omega}^{-1}X_*)^{-1}X_*'\hat{\Omega}^{-1}y_*$$

is consistent.

Proof: According to Proposition 14 of Section 2.5, the conclusions of Corollary 17 and Lemma 18 are sufficient to ensure that $\bar{\beta}_*$ is consistent.

20. LEMMA: $\text{plim}\,\dfrac{X_*'(\hat{\Omega}^{-1} - \Omega^{-1})\varepsilon_*}{\sqrt{T}} = 0$

Proof: The i-th subvector of the above vector is

$$\frac{1}{\sqrt{T}}\sum_{j=1}^{G} (s^{ij} - \sigma^{ij})X_i'\varepsilon_j$$

Its probability limit is

$$\sum_{j=1}^{G} \text{plim}(s^{ij} - \sigma^{ij})\text{plim}\,\frac{X_i'\varepsilon_j}{\sqrt{T}}$$

Now, according to Lemma 4 of Section 2.4, $X_i'\varepsilon_j/\sqrt{T}$ has the well-defined limiting distribution

$$N\left(0,\ \sigma_{jj}\lim_{T\to\infty}\frac{X_i'X_i}{T}\right)$$

Also, $\text{plim}(s^{ij} - \sigma^{ij}) = 0$. Therefore the above probability limit is zero, as required.

21. THEOREM (Zellner): The estimator

$$\bar{\beta}_* = (X_*'\hat{\Omega}^{-1}X_*)^{-1}X_*'\hat{\Omega}^{-1}y_*$$

has the same asymptotic distribution as the estimator

$$\tilde{\beta}_* = (X_*'\Omega^{-1}X_*)^{-1}X_*'\Omega^{-1}y_*$$

(Therefore $\bar{\beta}_*$ is asymptotically efficient.)

Proof: According to Theorem 15 of Section 2.5, the conclusions of Lemmas 16 and 20 are sufficient to ensure that $\bar{\beta}$ and $\tilde{\beta}$ have the same asymptotic distribution [see Note 1].

Concerning the small-sample properties of Zellner's "seemingly unrelated regressions" estimator $\bar{\beta}_*$, we can (except in certain special cases) only claim the following.

22. PROPOSITION: $\bar{\beta}_*$ is distributed symmetrically around β_*; it is unbiased if its mean exists.

Proof: According to Proposition 16 of Section 2.5, what we need to show is that $\hat{\Omega}$ is an even function of ε_*.

If ε_* changes sign, each of the OLS residual vectors e_i will change sign; however, $s_{ij} = e_i'e_j/T$ will be unaffected. Therefore,

$$\hat{\Omega} = S \otimes I$$

is unaffected by a sign change in ε_*; it is an even function of ε_*, which proves the proposition.

2.7 ASYMPTOTICALLY UNCOOPERATIVE REGRESSORS

This section will assume that the full ideal conditions hold except that the regressors may be asymptotically uncooperative, in the sense of the following definition.

1. DEFINITION: The regressor matrix X is said to be asymptotically uncooperative it if is not true that

$$Q = \lim_{T\to\infty} \frac{X'X}{T}$$

is finite and nonsingular.

A regressor can of course cause Q to be infinite if the regressor becomes "too big," or it can cause Q to be singular if it becomes "too small." For example, if one of the regressors is a time trend, some elements of Q will be infinite. If a regressor is of the form λ^t, where $0 < \lambda < 1$, then Q will be singular.

The point of this section is that it is really only the regressors which are "too small" which cause problems asymptotically.

2. PROPOSITION: A necessary and sufficient condition for the OLS estimator $\hat{\beta}$ to be consistent is that

$$\lim_{T\to\infty}(X'X)^{-1} = 0$$

Proof: $\hat{\beta}$ is normally distributed with mean β; it is therefore consistent if and only if its covariance matrix, $\sigma^2(X'X)^{-1}$, vanishes asymptotically.

3. LEMMA: Define d_{Ti} $(i = 1, 2, \cdots, K)$ by

$$d_{Ti}^2 = \sum_{t=1}^{T} X_{ti}^2$$

Define the matrix C by

$$C_{ij} = \lim_{T\to\infty} \frac{1}{d_{Ti}d_{Tj}} \sum_{t=1}^{T} X_{ti}X_{tj} \qquad (i = 1, 2, \cdots, K)$$

Then C is finite.

Proof: Note first that the diagonal elements of C equal one. Considering the off-diagonal elements, note that by the Schwarz inequality,

$$\left|\sum_{t=1}^{T}\left(\frac{X_{ti}}{d_{Ti}}\right)\left(\frac{X_{tj}}{d_{Tj}}\right)\right| \le \left\{\left[\sum_{T=1}^{T}\left(\frac{X_{ti}}{d_{Ti}}\right)^2\right]\left[\sum_{t=1}^{T}\left(\frac{X_{tj}}{d_{Tj}}\right)^2\right]\right\}^{\frac{1}{2}} = 1$$

Taking limits we get

$$|C_{ij}| \le 1$$

which is clearly finite.

4. THEOREM: Let d_{Ti} ($i = 1, 2, \cdots, K$) and C be as defined in Lemma 3. Then if

$$\text{(i)} \lim_{T\to\infty} d_{Ti}^2 = \infty$$

and

(ii) C is nonsingular

the OLS estimator $\hat{\beta}$ is consistent.

Proof: Let $X_* = XS$, where

$$S = \text{diagonal}\left(\frac{1}{d_{T1}}, \cdots, \frac{1}{d_{TK}}\right)$$

Note that

$$C = \lim_{T\to\infty} X_*'X_*$$

is finite (by Lemma 3) and nonsingular (by assumption). But we have

$$(X'X)^{-1} = S(X_*'X_*)^{-1}S$$

so

$$\lim_{T\to\infty}(X'X)^{-1} = \left(\lim_{T\to\infty} S\right)C^{-1}\left(\lim_{T\to\infty} S\right)$$

Since $\lim_{T\to\infty} S = 0$ if condition (i) is satisfied, we have $\lim_{T\to\infty}(X'X)^{-1} = 0$. By Proposition 2, this ensures that $\hat{\beta}$ is consistent.

Theorem 4 ought to be applicable in most cases where the regressors become "too big" for

$$\lim_{T\to\infty} \frac{X'X}{T}$$

to be finite. Essentially, it says that consistency is guaranteed as long as the sum of squares of each regressor approaches infinity with T.

The asymptotic distribution of the OLS estimator can also be established along the same lines.

5. PROPOSITION: Let C be as defined in Lemma 3, and S as defined in the proof of Theorem 4. Assume that C is nonsingular. Then

$$S^{-1}(\hat{\beta} - \beta)$$

has asymptotic distribution $N(0,\sigma^2 C^{-1})$.

Proof:

$$S^{-1}(\hat{\beta} - \beta) = S^{-1}(X'X)^{-1}X'\varepsilon$$
$$= S^{-1}S(X_*'X_*)^{-1}SX'\varepsilon$$

where, as before, $X_* = XS$. Now

$$\lim_{T\to\infty}(X_*'X_*)^{-1} = C^{-1}$$

Furthermore, $SX'\varepsilon$ is normally distributed with zero mean and with covariance matrix

$$\sigma^2 SX'XS = \sigma^2 X_*'X_*$$

Since $\lim_{T\to\infty} \sigma^2 X_*'X_* = C$, the asymptotic distribution of $SX'\varepsilon$ is $N(0,\sigma^2 C)$. Therefore the asymptotic distribution of $S^{-1}(\hat{\beta} - \beta)$ is $N(0,\sigma^2 C^{-1}\cdot C\cdot C^{-1}$ or simply $N(0,\sigma^2 C^{-1})$.

EXERCISES

1. Consider the linear model $y = X\beta + \varepsilon$, satisfying the ideal conditions except that $\varepsilon_t = \rho\varepsilon_{t-1} + u_t$, where the u_t are iid $N(0,\sigma^2)$, and where u_t is independent of ε_{t-i}, $i = 1, 2, \cdots, \infty$.
   a. Show that $\text{cov}(\varepsilon) = \sigma^2\Omega$, where $\Omega_{ij} = \rho^{|i-j|}$.
   b. Verify that $V'V = \Omega^{-1}$, where

$$V = (1 - \rho^2)^{-\frac{1}{2}} \begin{pmatrix} (1-\rho^2)^{\frac{1}{2}} & 0 & 0 & 0 & \cdots & 0 \\ -\rho & 1 & 0 & 0 & \cdots & 0 \\ 0 & -\rho & 1 & 0 & \cdots & 0 \\ 0 & 0 & -\rho & 1 & \cdots & 0 \\ \vdots & \vdots & \vdots & \vdots & \ddots & \vdots \\ 0 & 0 & 0 & 0 \cdots & -\rho & 1 \end{pmatrix}$$

c. Show that, except for the first observation, generalized least squares amounts to regressing $y_t - \rho y_{t-1}$ on $X_{ti} - \rho X_{t-1,i}$ $i = 1, 2, \cdots, K$.

d. Prove that the generalized least squares estimator of β based on any consistent estimate $\hat{\rho}$ has the same asymptotic distribution as the generalized least squares estimator of β based on the true value of ρ.

2. Find an example to show that the conditions of Proposition 14 and Theorem 15 of Section 2.5 are sufficient but *not necessary* for the conclusions of Proposition 14 and Theorem 15. One example is

$$\Omega^{-1} = \begin{pmatrix} 1 & \rho & \rho & \cdots & \rho \\ \rho & 1 & \rho & \cdots & \rho \\ \rho & \rho & 1 & \cdots & \rho \\ \vdots & \vdots & \vdots & & \vdots \\ \rho & \rho & \rho & \cdots & 1 \end{pmatrix} \qquad X = \begin{pmatrix} 1 \\ 1 \\ 1 \\ \vdots \\ 1 \end{pmatrix}$$

3. Consider a system of two seemingly unrelated regressions, in which the regressors for the second equation are a subset of the regressors for the first equation. Show that the OLS estimator of β_2 (the vector of coefficients of the second equation) is the same as the seemingly unrelated regressions estimator of β_2.

4. Consider a regression equation satisfying the ideal conditions except that one regressor (say the i-th) is of the form

$$X_{ti} = \lambda^t$$

where $|\lambda| < 1$. Show that the OLS estimator of β_i is inconsistent, but that the OLS estimates of the remaining coefficients are consistent.

5. Prove facts (a) and (b) given in Mathematical Review 5 of Section 2.3.

NOTES

Section 2.2

1. See the proof of Lemma 7 of Section 1.3.
2. $E[V'MV/(T - K)] = \sigma^2$ for exactly the same reason that $E[\epsilon'M\epsilon/(T - K)] = \sigma^2$ in the usual (classical) case. See Theorem 11 and Corollary 12 of Section 1.3.
3. This is just a reflection of the fact that

$$\begin{pmatrix} X_1'X_1 & X_1'X_2 \\ X_2'X_1 & X_2'X_2 \end{pmatrix}^{-1} \begin{pmatrix} X_1'X_1 \\ X_2'X_1 \end{pmatrix} = \begin{pmatrix} I \\ 0 \end{pmatrix}$$

4. This approach is due to Theil (1957).

Section 2.3

1. Identifiability will be discussed in Chapter 4.
2. This will generally be true; the exception would be the case in which a row of R happened to be a linear combination of the rows of X.

Section 2.4

1. The Chebychev inequality is sometimes used in an attempt to make nonparametric confidence statements about elements of β. Strictly speaking, this is not valid, since the variance σ^2 is unknown.
2. Compare Theorem 5 with Proposition 12 of Section 1.4.
3. For the details, see the proof of Lemma 2 of Section 1.4.
4. See the proof of Theorem 10 of Section 1.5.

Section 2.5

1. Some such definition must be made if one is to talk of the "consistency" of $\hat{\Omega}$ for Ω. The usual definition of consistency does not apply here since Ω is not of fixed dimension.
2. That is, a change in the sign of the disturbances would not change $\hat{\Omega}$.
3. This result is due to Kakwani (1967).

Section 2.6

1. This approach closely follows Theil (1971); for an alternative (direct) proof, see Zellner (1962).

REFERENCES

Amemiya, T. (1973), "Regression Analysis When the Variance of the Dependent Variable is Proportional to the Square of Its Expectation," *Journal of the American Statistical Association 68*, 928-934.

Anderson, T.W. (1971), *The Statistical Analysis of Time Series* New York: John Wiley.

Bock, M.E. (1975), "Minimax Estimators of the Mean of a Multivariate Normal Distribution," *Annals of Statistics 3*, 209-218.

Cochrane, D., and G.H. Orcutt (1949), "Application of Least Squares Regression to Relationships Containing Autocorrelated Error Terms," *Journal of the American Statistical Association 44*, 32-61.

Dhrymes, P.J. (1970), *Econometrics: Statistical Foundations and Applications* New York: Harper and Row, 1970.

Farebrother, R.W. (1975), "The Minimum Mean Square Error Linear Estimator and Ridge Regression," *Technometrics 17*, 127-128.

Goldberger, A.S. (1964), *Econometric Theory* New York: John Wiley.

Hoerl, A.E., and R.W. Kennard (1970), "Ridge Regression: Biased Estimation for Non-Orthogonal Problems," *Technometrics 12*, 35-67.

Johnston, J. (1972), *Econometric Methods*, 2nd ed. New York: McGraw Hill.

Kakawani, N.C. (1967), "The Unbiasedness of Zellner's Seemingly Unrelated Regression Equation Estimators," *Journal of the American Statistical Association 82*, 141-142.

Malinvand, E. (1966), *Statistical Methods of Econometrics* Chicago: Rand McNally.

Marquardt, D.W. (1970), "Generalized Inverses, Ridge Regression, Biased Linear Estimation, and Nonlinear Estimation," *Technometrics 12*, 591-612.

Rao, C.R. (1973), *Linear Statistical Inference and Its Application,* 2nd ed. New York: John Wiley.

Revankar, N.S. (1974), "Some Finite Sample Results in the Context of Two SUR Equations," *Journal of the American Statistical Association 69*, 187-190.

Theil, H. (1957), "Specification Errors and the Estimation of Economic Relationships," *Review of the International Statistical Institute 25*, 41-51.

———— (1971), *Principles of Econometrics* New York: John Wiley.

Zellner, A. (1962), "An Efficient Method of Estimating Seemingly Unrelated Regressions and Tests for Aggregation Bias," *Journal of the American Statistical Association 57*, 348-368.

CHAPTER 3

STOCHASTIC REGRESSORS

3.1 INTRODUCTION

This chapter will again consider the linear regression model $y = X\beta + \varepsilon$. It will be assumed that the full ideal conditions hold except that the regressor matrix X is random (at least in part) rather than nonstochastic. In particular, it is still assumed that $\varepsilon \sim N(0,\sigma^2 I)$, that

$$\underset{\sim}{Q} = \text{plim}\,\frac{X'X}{T}$$

is finite and nonsingular, and that the regressors (columns of X) are linearly independent with probability one; we relax *only* the assumption that X is nonstochastic.

Before proceeding to cases of specific interest, it is perhaps worth discussing (in a rather heuristic way) some general cases. First of all, consider the case in which X and ε are *independent*.

In this case the distribution of ε conditional on X is the same as its marginal distribution; specifically, conditional on X, $\varepsilon \sim N(0,\sigma^2 I)$

This immediately implies that, *conditional on X*, all the results of Chapter 1 hold. The OLS estimator has all the usual desirable properties, the usual tests are valid, and so forth. This is, of course, intuitively clear since, in making our statements conditional on X, we are essentially considering X as if it were nonstochastic; furthermore, since X and ε are independent, nothing is affected by the particular value of X on which our statements are conditional.

On the other hand, if both ε and X are stochastic, it is difficult to see the use of statements made conditional on X. Furthermore, it is very difficult to make inferences not conditional on X unless something is known about the distribution of X.

Consider, for example, the OLS estimator

$$\hat{\beta} = (X'X)^{-1}X'y = \beta + (X'X)^{-1}X'\varepsilon$$

Since $(X'X)^{-1}X'\varepsilon$ and $(X'X)^{-1}X'(-\varepsilon)$ have the same density, it is clear that the distribution of $\hat{\beta}$ is symmetric around β. Hence $E(\hat{\beta}) = \beta$ if $E(\hat{\beta})$ exists, *which it need not*. (Note that $y = X\beta + \varepsilon$ is not normally distributed, in general, unless X is; even in this case $\hat{\beta}$ is not normal.)

In order to consider the covariance matrix of $\hat{\beta}$, we will use the following lemma.

1. LEMMA: If $f(x,y)$ is the joint density of the random variables X,Y, and if $h(X,Y)$ is a function of X and Y such that $E[h(X,Y)]$ exists, then

$$E[h(X,Y)] = E_X\{E_{Y|X}[h(X,Y)]\}$$

where $E_{Y|X}$ is the expectation in the conditional distribution of Y given X and E_X is the expectation in the marginal distribution of X.

Proof: Factor $f(x,y)$ as $f(x,y) = k(y|x)g(x)$. Then

$$E[h(x,y)] = \int_{-\infty}^{\infty}\int_{-\infty}^{\infty} h(x,y)f(x,y)\ dx\ dy$$

$$= \int_{-\infty}^{\infty}\left(\int_{-\infty}^{\infty} h(x,y)k(y|x)dy\right)g(x)\ dx$$

$$= E_X\{E_{Y|X}[h(X,Y)]\}$$

If the covariance matrix of $\hat{\beta}$ exists (*it need not*), we therefore have

$$E(\hat{\beta} - \beta)(\hat{\beta} - \beta)' = E[(X'X)^{-1}X'\varepsilon\varepsilon'X(X'X)^{-1}]$$

$$= E[(X'X)^{-1}X'[E(\varepsilon\varepsilon'|X)]X(X'X)^{-1}]$$

$$= \sigma^2 E(X'X)^{-1}$$

This is clearly similar to the usual result $\sigma^2(X'X)^{-1}$.

Note, however, that the usual tests of hypotheses will not be valid in this case. For example, if we consider the usual t statistic used to test whether an element of β equals zero, the numerator does not have a normal distribution, the denominator is not the square root of a chi-squared distribution divided by its degrees of freedom, and the statistic does not have a t distribution. Rather, it has some distribution (presumably complicated) which must depend on the distribution of X. In the absence of more information about the distribution of X, not much is possible.

Similar comments apply asymptotically. It is sometimes argued that $\hat{\beta}$ can be expected to be consistent; since

$$\hat{\beta} = \beta + \left(\frac{X'X}{T}\right)^{-1} \frac{X'\varepsilon}{T}$$

and since we have assumed that

$$\underset{\sim}{Q} = \operatorname{plim} \frac{X'X}{T}$$

is finite and nonsingular, $\hat{\beta}$ will be consistent as long as

$$\operatorname{plim} \frac{X'\varepsilon}{T} = 0$$

This requirement is just that

$$\text{plim}\,\frac{1}{T}\sum_{t=1}^{T} X_{ti}\varepsilon_t = 0 \qquad (i = 1, 2, \cdots, K)$$

If the X_{ti}, $(t = 1, 2, \cdots, T)$ are iid, and if $E(X_{ti}\varepsilon_t)$ exists (in which case it equals zero), then the above result follows from the law of large numbers. However, the X_{ti}, $(t = 1, 2, \cdots, T)$ need not be iid, and $E(X_{ti}\varepsilon_t)$ need not exist, so that there is not even an assurance of consistency here [see Note 1].

In any case, we cannot be sure that $\hat{\beta}$ will be asymptotically normal; central limit theorems do not apply since $\hat{\beta}$ is a nonlinear combination of X and ε. (The nonlinearity is in the inversion of X'X.) Even asymptotically, valid tests of hypotheses are not possible, unless there is information about the distribution of X.

The above pessimistic comments concern the case in which X and ε are fully independent. When this is not the case, things only get worse. For example, if X_{ti} and ε_t are correlated, then typically

$$\text{plim}\,\frac{X'\varepsilon}{T} \neq 0$$

and $\hat{\beta}$ is inconsistent.

What we will attempt to cover in this chapter, therefore, are cases in which the regressors are stochastic for some particular reason, and where the particular nature of the regressors makes it possible to estimate and test hypotheses validly.

3.2 AUTOREGRESSIVE MODELS

Let us consider an autoregressive model of the form

$$y_t = \sum_{h=1}^{H} \gamma_h y_{t-h} + Z_t\alpha + \varepsilon_t$$

where the y_t are (scalar) observations on the dependent variable y, the γ_h are scalar parameters, Z_t is a $1 \times N$ vector of the t-th observations on each of N nonstochastic explanatory variables, α is the corresponding $N \times 1$ vector of parameters, and the ε_t are disturbances. The following assumptions are made:

(i) The ε_t are iid as $N(0,\sigma^2)$.

(ii) The elements Z_{ti} are uniformly bounded in absolute value; that is,

$$|Z_{ti}| < C$$

where $i = 1, 2, \cdots, N$, $t = 1, 2, \cdots, \infty$, and where C is a finite constant.

(iii) $\lim_{T\to\infty} \frac{1}{T - \Theta} \sum_{t=1}^{T-\Theta} Z_t' Z_{t+\Theta}$ exists for any integer Θ and is nonsingular for $\Theta = 0$.

(iv) The roots of the equation

$$\lambda^H - \sum_{h=1}^{H} \lambda^{H-h} \gamma_h = 0$$

are less than one in absolute value; that is, the model is dynamically stable.

Note that the above model can be written as

$$y_t = X_t \beta + \varepsilon_t$$

where

$$X_t = (y_{t-1}, \cdots, y_{t-H}, Z_t)$$

$$\beta = \begin{pmatrix} \gamma_1 \\ \vdots \\ \gamma_H \\ \alpha \end{pmatrix}$$

All T observations can be written in matrix form as simply

$$y = X\beta + \varepsilon$$

Note, however, that the total number of observations on y here is not just T, but T + H.

Properties of the Least Squares Estimator

1. COMMENT: Note that ε_t is independent of X_t, since X_t is made up of Z_t, which is nonstochastic, and of lagged values of y, which are (implicitly) composed only of lagged values of Z and of lagged disturbances. However, *all* observations on ε are not independent of *all* observations on X; for example, X_t contains y_{t-1} and is certainly not independent of ε_{t-1}. Hence, although $E(\varepsilon_t|X_t) = E(\varepsilon_t) = 0$, $E(\varepsilon|X) \neq 0$.

It is therefore clear that the OLS estimator of β will be biased. We have

$$\hat{\beta} = \beta + (X'X)^{-1}X'\varepsilon$$

and $E(X'X)^{-1}X'\varepsilon \neq 0$. Indeed, the small-sample distribution of $\hat{\beta}$ is quite intractable. Furthermore, note that this difficulty remains even if we consider the distribution of $\hat{\beta}$ conditional on X, since ε and X are not independent.

If we wish to find desirable properties for the OLS estimator, therefore, it must be asymptotic properties that we consider. Intuitively, since ε_t and X_t are independent, we might hope that

$$\text{plim}\ \frac{1}{T} X'\varepsilon = 0$$

and hence that $\hat{\beta}$ will be a consistent estimator of β. The following results show that this is indeed the case.

2. LEMMA: The matrix

$$\underline{Q} = \text{plim}\ \frac{X'X}{T}$$

is finite and nonsingular, and

$$\frac{X'\varepsilon}{\sqrt{T}}$$

converges in distribution to $N(0,\sigma^2\underline{Q})$.

Proof: The model considered here is exactly the model of Theorem 8 of Section A.3, corresponding there to G (the number of equations) = 1. The conclusion then follows immediately from Theorem 8, since the conditions of Theorem 8 are satisfied under the assumptions made above.

3. THEOREM: In the above autoregressive model, the OLS estimator $\hat{\beta} = (X'X)^{-1}X'y$ is consistent, and the asymptotic distribution of $\sqrt{T}(\hat{\beta} - \beta)$ is $N(0,\sigma^2 \underset{\sim}{Q}^{-1})$.

Proof: Since

$$\hat{\beta} = \beta + \left(\frac{X'X}{T}\right)^{-1} \frac{X'\varepsilon}{T}$$

then

$$\text{plim}\ \hat{\beta} = \beta + \underset{\sim}{Q}^{-1}\ \text{plim}\ \frac{X'\varepsilon}{T}$$

But since

$$\frac{X'\varepsilon}{\sqrt{T}}$$

has a well-defined asymptotic distribution, necessarily

$$\text{plim}\ \frac{X'\varepsilon}{T} = 0$$

Hence, plim $\hat{\beta} = \beta$.

Also,

$$\sqrt{T}(\hat{\beta} - \beta) = \left(\frac{X'X}{T}\right)^{-1} \frac{X'\varepsilon}{\sqrt{T}}$$

Since $X'\varepsilon/\sqrt{T}$ converges to $N(0,\sigma^2\underset{\sim}{Q})$, and since

$$\text{plim}\left(\frac{X'X}{T}\right)^{-1} = \underset{\sim}{Q}^{-1}$$

it follows that $\sqrt{T}(\hat{\beta} - \beta)$ converges in distribution to

$$N(0,\sigma^2\underset{\sim}{Q}^{-1}\underset{\sim}{Q}\ \underset{\sim}{Q}^{-1}) = N(0,\sigma^2\underset{\sim}{Q}^{-1})$$

4. COMMENT: This implies that the usual tests of hypotheses are *asymptotically* justified in this case.

5. PROPOSITION: The OLS estimator of β is the maximum likelihood estimator of β, conditional on $y_0, y_{-1}, \cdots, y_{-H+1}$.

Proof: The log likelihood function of $\varepsilon_1, \cdots, \varepsilon_T$ is of course

$$L = C - \frac{T}{2}\log\sigma^2 - \frac{1}{2}\sum_{t=1}^{T}\varepsilon_t^2$$

Since the Jacobian of the transformation from $\varepsilon_1, \varepsilon_2, \cdots, \varepsilon_T$ to $y_1, y_2, \cdots, y_T$ is one, we can write

$$L = C - \frac{T}{2}\log\sigma^2 - \frac{1}{2}\sum_{t=1}^{T}(y_t - X_t'\beta)^2$$

Maximizing this with respect to β does indeed lead to the OLS estimator $\hat{\beta} = (X'X)^{-1}X'y$.

Note, however, that T + H values of y are involved in the above expression. For example, X_1 contains $y_0, y_{-1}, \cdots, y_{-H+1}$. These are implicitly treated as fixed, that is, as given "initial conditions," and the likelihood function is conditional on these (fixed) values.

6. COMMENT: As long as the model is stable (as it is here by assumption), the influence of the initial conditions goes to zero asymptotically. Hence their treatment is of no asymptotic importance.

For example, in many cases one has the *same* number (T) of observations on y as on the fixed regressors Z. That is, the values $y_0, y_{-1}, \cdots, y_{-H+1}$ are not available. In this case a common procedure is to "drop" the first H observations, running the regression only on the last T - H observations. Another possible procedure is to assign arbitrary values (often zero) to $y_0, \cdots, y_{-H+1}$. Asymptotically it makes no difference which is done.

7. COMMENT: As a final note, it is perhaps worth commenting that autocorrelation [see Note 1] has far more serious consequences in this case than in the case in which all regressors are fixed. When X is fixed, the OLS estimator $\hat{\beta}$ is still consistent even in the presence of autocorrelation; this is *not* true in the present autoregressive model.

A standard example is the well-known geometric lag model

$$y_t = \beta \sum_{i=0}^{\infty} \lambda^i X_{t-i} + \varepsilon_t$$

Performing the "Koyck transformation" gives

$$y_t = \beta X_t + \lambda y_{t-1} + (\varepsilon_t - \lambda\varepsilon_{t-1})$$

Now, if X satisfies the conditions given earlier in this section, and if the model is stable ($|\lambda| < 1$), then

$$\text{plim } \frac{1}{T} \sum_{t=1}^{T} y_{t-1}(\varepsilon_t - \lambda\varepsilon_{t-1}) = -\lambda\sigma^2$$

The reason, of course, is that both y_{t-1} and $\varepsilon_t - \lambda\varepsilon_{t-1}$ contain ε_{t-1}. Hence the OLS estimator will be inconsistent.

3.3 INSTRUMENTAL VARIABLES

We again consider the model $y = X\beta + \varepsilon$. However, we will now be concerned with the case in which

$$\text{plim } \frac{1}{T} X'\varepsilon \neq 0$$

so that the OLS estimator $\hat{\beta} = (X'X)^{-1}X'y$ is inconsistent.

1. THEOREM: Suppose that there exists a set of variables Z_t such that

$$\underset{\sim}{Q}_{ZX} = \text{plim } \frac{1}{T} Z'X$$

is finite and nonsingular, and such that

$$\frac{Z'\varepsilon}{\sqrt{T}}$$

converges in distribution to $N(0,\psi)$. Then the estimator

$$\tilde{\beta} = (Z'X)^{-1}Z'y$$

is consistent, and the asymptotic distribution of $\sqrt{T}(\tilde{\beta} - \beta)$ is $N[0, \underset{\sim}{Q}_{ZX}^{-1} \psi(\underset{\sim}{Q}_{ZX}')^{-1}]$.

Proof: Note that

$$\tilde{\beta} = \beta + \left(\frac{Z'X}{T}\right)^{-1} \frac{Z'\varepsilon}{T}$$

Since

$$\frac{Z'\varepsilon}{\sqrt{T}}$$

has a well-defined asymptotic distribution, it follows that

$$\text{plim}\ \frac{Z'\varepsilon}{T} = 0$$

Hence plim $\tilde{\beta} = \beta$.

To prove the second part of the theorem, note that

$$\sqrt{T}(\tilde{\beta} - \beta) = \left(\frac{Z'X}{T}\right)^{-1} \frac{Z'\varepsilon}{\sqrt{T}}$$

The result follows immediately.

2. DEFINITION: The above estimator is the *instrumental variables* (IV) estimator of β; the matrix Z is the set of *instruments* for X.

3. COMMENT: The typical case is the one in which

$$\frac{Z'\varepsilon}{\sqrt{T}}$$

has asymptotic distribution $N(0,\sigma^2 Q_{ZZ})$, where

$$Q_{ZZ} = \text{plim}\ \frac{1}{T} Z'Z$$

4. EXAMPLE: As an example of the above technique, again consider the distributed lag model

$$y_t = \beta \sum_{i=0}^{\infty} \lambda^i X_{t-i} + \varepsilon_t$$

or

$$y_t = \beta X_t + \lambda y_{t-1} + (\varepsilon_t - \lambda\varepsilon_{t-1})$$

As a vector of instruments, one might choose $Z_t = (X_t, X_{t-1})$.

Under reasonable conditions on X, we have that

$$\text{plim}\ \frac{1}{T}\sum_{t=1}^{T}(X_t, X_{t-1})'(X_t, y_{t-1})$$

is finite and nonsingular, as required. Furthermore, since the disturbance $\varepsilon_t - \lambda\varepsilon_{t-1}$ is m-dependent (with m = 1), we can apply the central limit for m-dependent random variables (Theorem 7 of Section A.3) to show that

$$\frac{1}{\sqrt{T}}\sum_{t=1}^{T}(X_t, X_{t-1})'(\varepsilon_t - \lambda\varepsilon_{t-1})$$

has a well-defined asymptotic distribution. Hence the IV estimator will be consistent.

5. COMMENT: Theorem 1 above is stated somewhat differently from the "usual" treatment of instrumental variables estimators in the econometrics literature. For example, it is sometimes claimed that if $\underset{\sim}{Q}_{ZX}$ is finite and nonsingular and if

$$\text{plim}\ \frac{1}{T}Z'\varepsilon = 0$$

then the IV estimator $\tilde{\beta}$ is consistent, and $\sqrt{T}(\tilde{\beta} - \beta)$ converges in distribution to $N[0, \sigma^2 \underset{\sim}{Q}_{ZX}^{-1}\, \underset{\sim}{Q}_{ZZ}(\underset{\sim}{Q}_{ZX}^{-1})']$. The first part of this supposed theorem (consistency) is true, but the claim concerning the asymptotic distribution is *not true*. The reason is that

$$\text{plim}\ \frac{1}{T}Z'\varepsilon = 0$$

does not imply that

$$\frac{Z'\varepsilon}{\sqrt{T}}$$

converges in distribution to $N(0, \sigma^2 \underset{\sim}{Q}_{ZZ})$. As an example, consider the k-class estimator of a single structural equation (to be covered in Chapter 4), with

$$k = 1 + \frac{1}{\sqrt{T}}$$

The k-class estimator is an IV estimator (see Proposition 4 of Section 4.7). Furthermore, with the above value of k, we have $\underset{\sim}{Q}_{ZX}$ finite and nonsingular, $\underset{\sim}{Q}_{ZZ}$ finite, and

$$\text{plim} \frac{1}{T} Z'\varepsilon = 0$$

Yet $Z'\varepsilon/\sqrt{T}$ does not have a well-defined asymptotic distribution, and so neither does $\sqrt{T}(\tilde{\beta} - \beta)$. To put things another way, if the above false claim were true, it would imply that the k-class estimator with

$$k = 1 + \frac{1}{\sqrt{T}}$$

would have the same asymptotic covariance matrix as the two-stage least squares estimator. This is, of course, not true.

The above difficulty is unfortunate since it is clearly easier to verify a condition like

$$\text{plim} \frac{1}{T} Z'\varepsilon = 0$$

than it is to verify that

$$\frac{Z'\varepsilon}{\sqrt{T}}$$

has a well-defined asymptotic distribution. It is therefore obvious to inquire whether there is a useful, easily-verified sufficient condition for

$$\frac{Z'\varepsilon}{\sqrt{T}}$$

to have a well-defined asymptotic distribution. Of course, if Z is nonstochastic, and if

$$\underset{\sim}{Q}_{ZZ} = \text{plim} \frac{1}{T} Z'Z$$

is finite, this is sufficient. However, if Z is stochastic, things are rather complicated, and such sufficient conditions as might exist are apt to be too demanding to be of much use.

For example, one rather strong set of requirements is that the observations Z_t be iid and independent of all observations on ε [see Note 1]. But even this is not enough--consider, for example, the example of Section 3.1, where

$$Z_t = \frac{1}{X_t}$$

the X_t being iid $N(0,\sigma^2)$ and independent of ε. Clearly the Z_t are also independent of ε, and iid. Yet $Z_t\varepsilon_t$ has a Cauchy distribution (whose mean does not exist), and so does

$$\frac{1}{T}\sum_{t=1}^{T} Z_t\varepsilon_t$$

So it is not even true that

$$\text{plim}\,\frac{1}{T}\,Z'\varepsilon = 0$$

much less that

$$\frac{Z'\varepsilon}{\sqrt{T}}$$

has a well-defined asymptotic distribution.

3.4 ERRORS IN VARIABLES

Let us consider the simple linear regression model

$$y_t = \alpha + \beta X_t + \varepsilon_t$$

(X_t being a scalar). Suppose, however, that y_t and X_t are observed with error, so that the observed variables are

$$y_t^* = y_t + v_t$$

$$X_t^* = X_t + u_t$$

Note that we can then write the model as

$$y_t^* = \alpha + \beta X_t^* + (\varepsilon_t + v_t - \beta u_t)$$

1. THEOREM: Consider the model above, where it is assumed that the ε_t, v_t, u_t, and X_t are all independent of each other, and are furthermore iid as $N(0,\sigma_\varepsilon^2)$, $N(0,\sigma_v^2)$, $N(0,\sigma_u^2)$, and $N(\mu,\sigma_X^2)$, respectively. Then the OLS estimator $\hat{\beta}$ of β is inconsistent as long as $\sigma_u^2 \neq 0$;

$$\text{plim}\ \hat{\beta} = \beta \frac{\sigma_X^2}{\sigma_X^2 + \sigma_u^2} \le \beta$$

Proof: The OLS estimator of β is

$$\hat{\beta} = \frac{\sum_t (X_t^* - \bar{X}^*)(y_t^* - \bar{y}^*)}{\sum_t (X_t^* - \bar{X}^*)^2}$$

$$= \beta + \frac{(1/T)\sum_t (X_t^* - \bar{X}^*)(\varepsilon_t + v_t - \beta u_t)}{(1/T)\sum_t (X_t^* - \bar{X}^*)^2}$$

Since $X_t^* = X_t + u_t$, under the assumptions made above we have

$$\text{plim}\ \frac{1}{T} \sum_t (X_t^* - \bar{X}^*)(\varepsilon_t + v_t - \beta u_t) = -\beta\sigma_u^2$$

and

$$\text{plim}\ \frac{1}{T} \sum_t (X_t^* - \bar{X}^*)^2 = \sigma_X^2 + \sigma_u^2$$

Hence

$$\text{plim}\ \hat{\beta} = \beta + \frac{-\beta\sigma_u^2}{\sigma_X^2 + \sigma_u^2}$$

$$= \beta \frac{\sigma_X^2}{\sigma_X^2 + \sigma_u^2} \le \beta$$

2. COMMENT: It should be clear that it is the error in measuring X which causes the problem here. Indeed, the error v in measuring y is indistinguishable from the usual disturbance ε. For the remainder of this section we will therefore dispense with the disturbance ε, incorporating it into the measurement error v.

Theorem 1 has shown that OLS is inconsistent in the above context. The following result shows that *no* estimation procedure could hope to be consistent.

3. THEOREM: Suppose that the conditions of Theorem 1 hold (except as noted in Comment 2). Then the only parameter which is identified is $\mu = E(X)$.

Proof: Under the conditions of Theorem 1, the distribution of the observable quantities y^* and X^* is normal and is characterized by the quantities

$$\begin{aligned} E(X^*) &= \mu \\ E(y^*) &= \alpha + \beta\mu \\ \mathrm{Var}(X^*) &= \sigma_X^2 + \sigma_u^2 \\ \mathrm{Var}(y^*) &= \beta^2\sigma_X^2 + \sigma_v^2 \\ \mathrm{Cov}(X^*,Y^*) &= \beta\sigma_X^2 \end{aligned}$$

Clearly, these two expectations, two variances, and one covariance are all that any set of observations on X^* and y^* can hope to reveal.

Note, however, that there are *six* unknown parameters: α, β, μ, σ_X^2, σ_u^2, σ_v^2. Given any set of values of these parameters, we can find an infinite number of other sets of values that would give the same distribution of X^* and y^*; the parameters are not all identified. Only $\mu = E(X) = E(X^*)$ is uniquely determined from knowledge of the distribution of X^* and y^*.

4. COMMENT: The assumptions of the above discussion have been slightly different from the assumptions which usually underlie the regression model, in that X_t has been assumed to be distributed as $N(\mu,\sigma_X^2)$. It would be more in accord with our usual assumptions to treat the X_t as nonstochastic, though of course changing with t. But this, if anything, only makes matters worse, since we then change the mean of X_t^* to X_t and the mean of y_t^* to $\alpha + \beta X_t$; we essentially replace the one unknown (though identified) parameter μ with T unknown parameters $X_1, X_2, \cdots, X_T$. Clearly we now have far more parameters than we can hope to estimate; indeed, *no* parameters are now identified.

5. PROPOSITION: As above, let

$$y_t = \alpha + \beta X_t$$

$$y_t^* = y_t + v_t$$

$$X_t^* = X_t + u_t$$

Assume that the X_t, v_t, and u_t are mutually independent, that the v_t and u_t are iid as $N(0,\sigma_v^2)$ and $N(0,\sigma_u^2)$, respectively, and that the X_t are also iid. Then if β is not identified, X_t is either normally distributed or constant.

Proof: The joint distribution of u and v is normal; so the characteristic function of the distribution of u and v is

$$\phi_{uv}(t_1,t_2) = \exp[-\tfrac{1}{2}(\sigma_u^2 t_1^2 + \sigma_v^2 t_2^2)]$$

(The notation ϕ_{uv} is used to emphasize that this is the characteristi function of the joint distribution of u and v.) Also the characteristic function of X and y is

$$\begin{aligned}\phi_{Xy}(t_1,t_2) &= E[\exp(iXt_1 + iyt_2)]\\ &= E[\exp(iXt_1 + i\alpha t_2 + i\beta Xt_2)]\\ &= \exp[i\alpha t_2]\phi_X(t_1 + \beta t_2)\end{aligned}$$

Finally, since (X,y) and (u,v) are independent

$$\begin{aligned}\phi_{X^*y^*}(t_1,t_2) &= \phi_{Xy}(t_1,t_2) \cdot \phi_{uv}(t_1,t_2)\\ &= \exp[i\alpha t_2 - \tfrac{1}{2}(\sigma_u^2 t_1^2 + \sigma_v^2 t_2^2)]\phi_X(t_1 + \beta t_2)\end{aligned}$$

Now, let us assume that we have alternative structures

$$\{\alpha,\beta,\sigma_u^2,\sigma_v^2,\phi_X\}$$

and

$$\{\alpha^o,\beta^o,\sigma_u^{2\,o},\sigma_v^{2\,o},\phi_X^o\}$$

which imply the same distribution for (X^*,y^*). Hence

$$\exp[i\alpha t_2 - \frac{1}{2}(\sigma_u^2 t_1^2 + \sigma_v^2 t_2^2)]\phi_X(t_1 + \beta t_2)$$
$$= \exp[i\alpha^o t_2 - \frac{1}{2}(\sigma_u^{2\,o} t_1^2 + \sigma_v^{2\,o} t_2^2)]\phi_X^o(t_1 + \beta^o t_2)$$

for all t_1 and t_2. Furthermore, if β is not identified, we can pick these structures such that $\beta \neq \beta^o$. This in turn implies that for any $a \neq 0$, there exist t_1 and t_2 such that

$$t_1 + \beta t_2 = a$$
$$t_1 + \beta^o t_2 = 0$$

explicitly,

$$t_1 = \frac{-\beta^o a}{\beta - \beta^o}$$
$$t_2 = \frac{a}{\beta - \beta^o}$$

Substituting above, we have

$$\exp\left\{\frac{i\alpha a}{\beta - \beta_o} - \frac{1}{2}\left[\frac{\sigma_u^2 \beta^{o\,2} a^2}{(\beta - \beta^o)^2} + \frac{\sigma_v^2 a^2}{(\beta - \beta^o)^2}\right]\right\}\phi_X(a)$$

$$= \exp\left\{\frac{i\alpha^o a}{\beta - \beta_o} - \frac{1}{2}\left[\frac{\sigma_u^{2\,o} \beta^{o2} a^2}{(\beta - \beta^o)^2} + \frac{\sigma_v^{2\,o} a^2}{(\beta - \beta^o)^2}\right]\right\}\phi_X^o(0)$$

But $\phi_X^o(0) = 1$, and so this gives

$$\phi_X(a) = \exp\left\{ i\,\frac{\alpha^o - \alpha}{\beta - \beta^o}\,a - \frac{1}{2(\beta - \beta^o)^2}\left[(\sigma_u^{2\,o} - \sigma_u^2)\beta^{o\,2} + (\sigma_v^{2\,o} - \sigma_v^2)\right]a^2\right\}$$

Finally, this is the characteristic function of a normal random variable, or, if $\sigma_u^{2\,o} - \sigma_u^2 = \sigma_v^{2\,o} - \sigma_v^2 = 0$, a constant.

6. COMMENT: From Theorem 3 and Proposition 5, it is clear that identification problems arise when u, v, and X are *all* normal, that

is, when *both* y* and X* are normal. Under this assumption, therefore, we can not hope to find consistent estimates unless we utilize some further a priori information. Alternatively, we would consider estimators which do not utilize prior information, realizing that these can only be consistent for (possibly) certain types of nonnormal regressors or errors.

Let us first consider estimation using additional prior information. Under the assumptions given above, we see that we have a random sample of size T on (X*,y*), which is normally distributed with

$$
\begin{aligned}
E(X^*) &= \mu \\
E(y^*) &= \alpha + \beta\mu \\
\sigma_{X^*}^2 &= \sigma_X^2 + \sigma_u^2 \\
\sigma_{y^*}^2 &= \beta^2\sigma_X^2 + \sigma_v^2 \\
\operatorname{cov}(X^*, y^*) &= \beta\sigma_X^2
\end{aligned}
$$

as noted in the proof of Theorem 3. Hence we can obtain maximum likelihood estimates

$$
\begin{aligned}
\hat{E}(X^*) &= \bar{X}^* \\
\hat{E}(y^*) &= \bar{y}^* \\
\hat{\sigma}_{X^*}^2 &= \frac{1}{T}\sum_{t=1}^{T}(X_t^* - \bar{X}^*)^2 \\
\hat{\sigma}_{y^*}^2 &= \frac{1}{T}\sum_{t=1}^{T}(y_t^* - \bar{y}^*)^2 \\
\hat{\operatorname{cov}}(X^*, y^*) &= \frac{1}{T}\sum_{t=1}^{T}(X_t^* - \bar{X}^*)(y_t^* - \bar{y}^*)
\end{aligned}
$$

and we obtain maximum likelihood estimates of α, β, μ, σ_X^2, σ_u^2, σ_v^2, by solving the set of equations

$$\hat{E}(X^*) = \mu$$

$$\hat{E}(y^*) = \alpha + \beta\mu$$

$$\hat{\sigma}^2_{X^*} = \sigma^2_X + \sigma^2_u$$

$$\hat{\sigma}^2_{y^*} = \beta^2\sigma^2_X + \sigma^2_v$$

$$\hat{\text{cov}}(X^*,y^*) = \beta\sigma^2_X$$

Clearly we have $\hat{\mu} = \hat{E}(X^*) = \overline{X}^*$ and, if β is identified, $\hat{\alpha} = \hat{E}(y^*) - \hat{\beta}\hat{\mu} = \overline{y}^* - \hat{\beta}\overline{X}^*$. Eliminating these two equations and variables, we are left with

$$\hat{\sigma}^2_{X^*} = \sigma^2_X + \sigma^2_u$$

$$\hat{\sigma}^2_{y^*} = \beta^2\sigma^2_X + \sigma^2_v$$

$$\hat{\text{cov}}(X^*,y^*) = \beta\sigma^2_X$$

This is a set of three equations in four unknowns (β, σ^2_X, σ^2_u, σ^2_v), and so, as we know from Theorem 3, estimation is impossible without prior information. The following result shows how such prior information may be utilized.

7. PROPOSITION: When σ^2_v is known, the maximum likelihood estimate of β is

$$\hat{\beta} = \frac{\sum_{t=1}^{T}(y^*_t - \overline{y}^*)^2 - \sigma^2_v}{\sum_{t=1}^{T}(X^*_t - \overline{X}^*)(y^*_t - \overline{y}^*)}$$

When σ^2_u is known, the maximum likelihood estimate of β is

$$\hat{\beta} = \frac{\sum_{t=1}^{T}(X^*_t - \overline{X}^*)(y^*_t - \overline{y}^*)}{\sum_{t=1}^{T}(X^*_t - \overline{X}^*)^2 - \sigma^2_u}$$

When $\lambda = \sigma^2_v/\sigma^2_u$ is known, the maximum likelihood estimate of β is

$$\frac{\sum_{t=1}^{T}(y_t^* - \bar{y}^*)^2 - \lambda \sum_{t=1}^{T}(X_t^* - \bar{X}^*)^2}{2\sum_{t=1}^{T}(X_t^* - \bar{X}^*)(y_t^* - \bar{y}^*)}$$

$$+\sqrt{\frac{[\sum_{t=1}^{T}(y_t^* - \bar{y}^*)^2 - \lambda \sum_{t=1}^{T}(X_t^* - \bar{X}^*)^2]^2 + 4\lambda[\sum_{t=1}^{T}(X_t^* - \bar{X}^*)(y_t^* - \bar{y}^*)]^2}{2\sum_{t=1}^{T}(X_t^* - \bar{X}^*)(y_t^* - \bar{y}^*)}}$$

Proof: Suppose first that σ_v^2 is known. Then from the second and third of the three equations in the discussion preceeding the proposition, we have

$$\hat{\beta} = \frac{\hat{\sigma}_{y^*}^2 - \sigma_v^2}{\hat{\text{cov}}(X^*,y^*)}$$

This is just the result claimed.

Similarly, when σ_u^2 is known, from the first and third of the above equations we have

$$\hat{\beta} \quad \frac{\hat{\text{cov}}(X^*,y^*)}{\hat{\sigma}_{X^*}^2 - \sigma_u^2}$$

as claimed.

Finally, consider the case in which $\lambda = \sigma_v^2/\sigma_u^2$ is known. We have

$$\sigma_u^2 = \hat{\sigma}_{X^*}^2 - \sigma_X^2 = \hat{\sigma}_{X^*}^2 - \frac{\hat{\text{cov}}(X^*,y^*)}{\hat{\beta}}$$

and

$$\sigma_v^2 = \hat{\sigma}_{y^*}^2 - \hat{\beta}^2\sigma_X^2 = \hat{\sigma}_{y^*}^2 - \hat{\beta}\,\hat{\text{cov}}(X^*,y^*)$$

Hence

$$\lambda = \frac{\sigma_v^2}{\sigma_u^2} = \frac{\hat{\beta}\hat{\sigma}_{y^*}^2 - \hat{\beta}^2\,\hat{\text{cov}}(X^*,y^*)}{\hat{\beta}\hat{\sigma}_{X^*}^2 - \hat{\text{cov}}(X^*,y^*)}$$

which yields the quadratic equation

$$\hat{\beta}^2\,\hat{\text{cov}}(X^*,y^*) + \hat{\beta}(\lambda\hat{\sigma}_{X^*}^2 - \hat{\sigma}_{y^*}^2) - \lambda\,\hat{\text{cov}}(X^*,y^*) = 0$$

The solution is

$$\hat{\beta} = \frac{(\hat{\sigma}_{y*}^2 - \lambda\hat{\sigma}_{X*}^2) \pm \sqrt{(\lambda\hat{\sigma}_{X*}^2 - \hat{\sigma}_{y*}^2)^2 + 4\lambda\, \hat{\text{cov}}^2(X^*,y^*)}}{2\, \hat{\text{cov}}(X^*,y^*)}$$

which is the result claimed. Note that the positive square root is used so that $\hat{\beta}$ has the same sign as $\hat{\text{cov}}(X^*,y^*)$.

8. COMMENT: Recall that we have incorporated into v the measurement error on y *plus* the disturbance term. The variance of the disturbance term is not apt to be known a priori, which implies that prior information on σ_v^2 or $\lambda = \sigma_v^2/\sigma_u^2$ is not apt to be forthcoming. On the other hand, σ_u^2 (the variance of the measurement error on X) may sometimes be known, at least approximately, if it is known how the data were obtained.

Let us now consider estimation which does not incorporate prior information. The best-known such procedure is that based on grouping of the observations.

9. PROPOSITION: Suppose that the T observations on (X^*,y^*) are placed into (at least) two groups. Let G_1 represent the first group, with T_1 observations, and G_2 represent the second group, with T_2 observations. Define

$$\hat{b}_1 = \frac{1}{T_1}\sum_{t\in G_1} y_t^* - \frac{1}{T_2}\sum_{t\in G_2} y_t^*$$

$$\hat{b}_2 = \frac{1}{T_1}\sum_{t\in G_1} X_t^* - \frac{1}{T_2}\sum_{t\in G_2} X_t^*$$

$$\hat{\beta} = \frac{\hat{b}_1}{\hat{b}_2}$$

Then $\hat{\beta}$ is a consistent estimator of β if the grouping is independent of the measurement errors (u and v) and if plim $\hat{b}_2 \neq 0$.

Proof: Note that if X and y were observable, we could define

$$b_2 = \frac{1}{T_1}\sum_{t\epsilon G_1} X_t - \frac{1}{T_2}\sum_{t\epsilon G_2} X_t$$

$$b_1 = \frac{1}{T_1}\sum_{t\epsilon G_1} y_t - \frac{1}{T_2}\sum_{t\epsilon G_2} y_t = \beta b_2$$

Hence $\beta = b_1/b_2$. This is the motivation for the estimator.

In the present case,

$$\hat{b}_1 - b_1 = \frac{1}{T_2}\sum_{t\epsilon G_1} v_t - \frac{1}{T_2}\sum_{t\epsilon G_2} v_t$$

Now, if the grouping is independent of v, both terms have a probability limit of zero, so that plim $\hat{b}_1$ = plim b_1. Similarly, if the grouping is independent of u, plim $\hat{b}_2$ = plim b_2. Hence

$$\text{plim}\ \hat{\beta} = \frac{\text{plim}\ \hat{b}_1}{\text{plim}\ \hat{b}_2} = \text{plim}\ \frac{b_1}{b_2} = \beta$$

if plim $\hat{b}_2 \neq 0$.

10. COMMENT: The difficulty with the above procedure is that if the grouping is independent of X, as well as u, then

$$\text{plim}\ \frac{1}{T_1}\sum_{t\epsilon T_1} X_t^* = \text{plim}\ \frac{1}{T_2}\sum_{t\epsilon T_2} X_t^* = \mu$$

so that plim $\hat{b}_2$ = 0 and the above proof fails. Hence the problem is to find a grouping which depends on X but not on u. Since all that is observed is $X^* = X + u$, this is clearly not a trivial problem. Indeed, in general no such grouping is possible, as should be clear from our discussion of identification above.

Wald (1940) has suggested grouping the data according to the largest and smallest values of X_t^*. Unfortunately this will generally depend on u. One exception is if there can be no observations X_t^* in the range $(x - c, x + c)$, where x is the median of X^* and c is the maximum range of the error u. Clearly this requires u to have a finite range and X to have a strange distribution as well.

Other groupings based on X^* have been suggested. All suffer from this same problem.

As a closing note, we will lastly consider one case in which the regressor is measured with error and yet none of the above problems arise. Suppose that instead of assuming

$$X^* = X + u$$

with u uncorrelated with X but correlated with X^*, we assume

$$X = X^* + u$$

with u uncorrelated with X^* but correlated with X. Such a case might typically arise in a controlled experiment. For example, a physicist might set an ohmmeter at 10 ohms, thus observing $X^* = 10$. The true resistance is then different from 10 by some random error which would then be correlated with the true value X but not with the (fixed) ohmmeter setting X^*.

In any case, if the model is

$$y = \alpha + \beta X$$
$$X = X^* + u$$
$$y^* = y + v$$

we have

$$y^* = \alpha + \beta X^* + (v + \beta u)$$

If u is independent of X^* (and v is too), then OLS will provide a consistent estimate of β in general.

Such a case is not apt to be common in economics, since economic data are not typically generated in controlled experiments.

EXERCISES

1. Find the asymptotic distribution of the estimator given in Example 4 of Section 3.3.

2. Consider an autoregressive model with disturbance following a first-order autoregressive process: $\varepsilon_t = \rho\varepsilon_{t-1} + u_t$, where $|\rho| < 1$, the u's are iid as $N(0,\sigma^2)$, and $\text{cov}(u_t\varepsilon_{t-i}) = 0$, $i = 1, 2, \cdots, K$.

a. Show that the OLS estimator is inconsistent.

b. Find a consistent instrumental variables estimator.

c. Show that the generalized least squares estimator based on a consistent estimate of ρ does not have the same asymptotic distribution as the generalized least squares estimator based on ρ itself.

3. Show that the "grouping" estimator given in Proposition 9 of Section 3.4 is an instrumental variables estimator.

NOTES

Section 3.1

1. For example, let $X_{t1} = 1/w_t$, where the w_t are iid as $N(0,\sigma^2)$ and are independent of ε_t. Then $X_{t1}\varepsilon_t$ has a Cauchy distribution, and

$$\text{plim}\ \frac{1}{T}\sum_{t=1}^{T} X_{t1}\varepsilon_t$$

does not even exist, much less equal zero.

Section 3.2

1. Autocorrelation is defined as the case in which

$$\text{cov}(\varepsilon_t,\varepsilon_s) \neq 0$$

for some $t \neq s$.

Section 3.3

1. This is essentially the case considered by Sargan (1958).

REFERENCES

Berkson, J. (1950), "Are There Two Regressions?," *Journal of the American Statistical Association* 45, 164-180.

Goldberger, A.S. (1964), *Econometric Theory* New York: John Wiley.

Johnston, J. (1972), *Econometric Methods*, 2nd ed. New York: McGraw Hill.

Madansky, A. (1959), "The Fitting of Straight Lines When Both Variables Are Subject to Error," *Journal of the American Statistical Association 54*, 163-205.

Malinvaud, E. (1966), *Statistical Methods of Econometrics* Chicago: Rand McNally.

Mann, H.B., and A. Wald (1943), "On the Statistical Treatment of Linear Difference Equations," *Econometrica 11*, 173-220.

Riersol, O. (1950), "Identifiability of a Linear Relation Between Variables Which Are Subject to Error," *Econometrica 18*, 375-389.

Sargan, J.D. (1958), "The Estimation of Economic Relationships Using Instrumental Variables," *Econometrica 26*, 393-415.

Theil, H. (1971), *Principles of Econometrics* New York: John Wiley.

Wald, A. (1940), "The Fitting of Straight Lines if Both Variables Are Subject to Error," *Annals of Mathematical Statistics 11*, 284-300.

CHAPTER 4

SIMULTANEOUS EQUATIONS I

4.1 INTRODUCTION

Let us consider a set of G simultaneous equations:

$$\sum_{j=1}^{G} Y_{tj}\Gamma_{ji} + \sum_{j=1}^{K} X_{tj}\Delta_{ji} + \varepsilon_{ti} = 0 \qquad (i = 1, 2, \cdots, G \text{ and } t = 1, 2, \cdots, T)$$

Note that there are T observations; G equations determining the G endogenous variables $Y_{t1}, \cdots, Y_{tG}$; K predetermined variables $X_{t1}, \cdots, X_{tK}$; and a disturbance for each equation. It is assumed that the predetermined variables are either nonstochastic (exogenous) or lagged endogenous variables.

The model can be written in matrix form as

$$Y\Gamma + X\Delta + \varepsilon = 0$$

Clearly the dimensions of Y, Γ, X, Δ, and ε are $T \times G$, $G \times G$, $T \times K$, $K \times G$, and $T \times G$, respectively. One observation (say the t-th) on all G equations can be written as

$$Y_{t\cdot}\Gamma + X_{t\cdot}\Delta + \varepsilon_{t\cdot} = 0$$

where $Y_{t\cdot}$, $X_{t\cdot}$, and $\varepsilon_{t\cdot}$ are the t-th rows of Y, X, and ε. Similarly, all observations on any one equation (say the i-th) can be written as

$$Y\Gamma_{\cdot i} + X\Delta_{\cdot i} + \varepsilon_{\cdot i} = 0$$

where $\Gamma_{\cdot i}$, $\Delta_{\cdot i}$, and $\varepsilon_{\cdot i}$ are the i-th columns of Γ, Δ, and ε.

The statistical content of the model is embodied in the following assumption.

1. ASSUMPTION:

(i) The vectors $\varepsilon'_{t\cdot}$ are independently and identically distributed as $N(0,\Sigma)$.

(ii) Γ is nonsingular.

(iii) Let W be the submatrix of X corresponding to the purely exogenous (nonstochastic) variables. Then

$$\underset{\sim}{Q}^* = \lim_{T\to\infty} \frac{1}{T} W'W = \lim_{T\to\infty} \frac{1}{T} \sum_{t=1}^{T} w'_{t\cdot}w_{t\cdot}$$

is finite and nonsingular. Also,

$$\underset{\sim}{Q}^{(k)} = \lim_{T\to\infty} \frac{1}{T-k} \sum_{t=1}^{T-k} w'_{t\cdot}w_{t+k\cdot}$$

is finite for $k = 1, 2, \cdots$.

(iv) Again distinguishing the purely exogenous variables in X from the lagged endogenous variables, write $X\Delta$ as

$$X\Delta = W\Delta_{(0)} + Y_{(-1)}\Delta_{(1)} + \cdots + Y_{(-H)}\Delta_{(H)}$$

Here $Y_{(-i)}$ is the $T \times G$ matrix of observations on Y, lagged i periods, and $\Delta_{(i)}$ is the corresponding $G \times G$ matrix of coefficients; H is the maximum lag encountered. Then it is assumed that all roots of the determinental equation

$$|\lambda^{H}\Gamma + \lambda^{H-1}\Delta_{(1)} + \cdots + \Delta_{(H)}| = 0$$

are less than one in absolute value.

(v) The elements of W are uniformly bounded (in absolute value).

2. COMMENT: Part (i) of Assumption 1 implies that

$$E\,\frac{\varepsilon'\varepsilon}{T} = \text{plim}\,\frac{\varepsilon'\varepsilon}{T} = \Sigma$$

4.2 THE REDUCED FORM

The set of equations just considered is a set of G equations in G endogenous variables, plus predetermined variables and disturbances. Given that Γ^{-1} exists (as assumed), we can solve for the endogenous variables in terms of the predetermined variables and disturbances. To do so, simply multiply the equation $Y\Gamma + X\Delta + \varepsilon = 0$ by Γ^{-1} to get

$$Y = -X\Delta\Gamma^{-1} - \varepsilon\Gamma^{-1}$$

1. DEFINITION: The *reduced form* of the model $Y\Gamma + X\Delta + \varepsilon = 0$ is

$$Y = X\Pi + V$$

where $\Pi = -\Delta\Gamma^{-1}$, $V = -\varepsilon\Gamma^{-1}$.

Note that the dimensions of Π and V are $K \times G$ and $T \times G$, respectively.

2. PROPOSITION: Let $V_{t\cdot}$ be the t-th row of $V = -\varepsilon\Gamma^{-1}$. Then the vectors $V'_{t\cdot}$ are independently and identically distributed as $N(0,\Omega)$, where

$$\Omega = (\Gamma^{-1})'\Sigma\Gamma^{-1}$$

Proof: By definition, $V'_{t\cdot} = -(\Gamma^{-1})'\varepsilon'_{t\cdot}$. Since the $\varepsilon'_{t\cdot}$ are iid as $N(0,\Sigma)$, the $V'_{t\cdot}$ are also iid as normal with zero mean and covariance matrix

$$(\Gamma^{-1})'\Sigma(-\Gamma^{-1}) = \Omega$$

4.3 ORDINARY LEAST SQUARES

1. PROPOSITION:

$$\underset{\sim}{Q} = \text{plim}\,\frac{X'X}{T}$$

is finite and nonsingular; furthermore,

$$\text{vec}\,\frac{X'V}{\sqrt{T}} = \frac{1}{\sqrt{T}}\begin{pmatrix} X'V_{\cdot 1} \\ \vdots \\ X'V_{\cdot G} \end{pmatrix} \qquad \text{[see Note 1]}$$

has asymptotic distribution $N(0, \Omega \otimes \underset{\sim}{Q})$.

Proof: The reduced-form equation

$$Y_{t\cdot} = X_{t\cdot}\Pi + V_{t\cdot}$$

can be written as

$$Y_{t\cdot} = W_t\Pi_{(0)} + Y_{t-1,\cdot}\Pi_{(1)} + \cdots + Y_{t-H,\cdot}\Pi_{(H)} + V_{t\cdot}$$

corresponding to the partition of X as in Assumption 1 of Section 4.1. Note that if the matrices $\Delta_{(i)}$ are as defined there, then

$$\Pi_{(i)} = -\Delta_{(i)}\Gamma^{-1} \qquad (i = 0, 1, \cdots, H)$$

This is a multivariate autoregressive process. Theorem 8 of Section A.3 then gives the desired conclusion, given that its conditions are met, which we now check.

Condition (i) of Theorem 8 is met since, by Proposition 2 of Section 4.2, the disturbances $V_{t\cdot}'$ are iid as $N(0,\Omega)$. Also requirements (ii) and (iii) of Theorem 8 are satisfied; see Assumption 1 of Section 4.1. This leaves only requirement (iv) of Theorem 8 to be checked. In the present context this requirement is just that the roots of

$$\left|\lambda^H I - \sum_{h=1}^{H} \lambda^{H-h}\Pi_{(h)}\right| = 0$$

be less than one in absolute value. But, multiplying through by $|\Gamma|$, this is equivalent to requiring that the roots of

$$\left|\lambda^H\Gamma + \sum_{h=1}^{H} \lambda^{H-h}\Delta_{(h)}\right| = 0$$

be less than one in absolute value. And this is guaranteed by Assumption 1 of Section 4.1.

2. THEOREM:

$$\text{vec}\,\frac{X'\varepsilon}{\sqrt{T}}$$

has asymptotic distribution $N(0, \Sigma \otimes \underset{\sim}{Q})$.

Proof: We know that

$$\text{vec}\,\frac{X'V}{\sqrt{T}}$$

has asymptotic distribution $N(0, \Omega \otimes \underset{\sim}{Q})$. Also we know that $\varepsilon = -V\Gamma$. Hence

$$\text{vec}\,\frac{X'\varepsilon}{\sqrt{T}} = \text{vec}\,\frac{X'V\Gamma}{\sqrt{T}} = (\Gamma' \otimes I)\,\text{vec}\,\frac{X'V}{\sqrt{T}} \qquad \text{[see Note 2]}$$

has asymptotic distribution

$$N(0, \Gamma'\Omega\Gamma \otimes \underset{\sim}{Q}) = N(0, \Sigma \otimes \underset{\sim}{Q})$$

3. COROLLARY: $\text{plim}\, X'\varepsilon/T = 0$.

4. COROLLARY: $\text{plim}\, X'V/T = 0$.

5. PROPOSITION:

$$\text{plim}\,\frac{1}{T}\begin{pmatrix} X'X & X'Y & X'\varepsilon \\ Y'X & Y'Y & Y'\varepsilon \\ \varepsilon'X & \varepsilon'Y & \varepsilon'\varepsilon \end{pmatrix} = \begin{pmatrix} \underset{\sim}{Q} & \underset{\sim}{Q}\Pi & 0 \\ \Pi'\underset{\sim}{Q} & \Pi'\underset{\sim}{Q}\Pi + \Omega & -(\Gamma^{-1})'\Sigma \\ 0 & -\Sigma\Gamma^{-1} & \Sigma \end{pmatrix}$$

Proof: We already know that

$$\text{plim}\,\frac{1}{T}\,X'X = \underset{\sim}{Q}$$

$$\text{plim}\,\frac{1}{T}\,X'\varepsilon = 0$$

$$\text{plim}\,\frac{1}{T}\,\varepsilon'\varepsilon = \Sigma$$

Now

$$\text{plim}\,\frac{1}{T}\,X'Y = \text{plim}\,\frac{1}{T}\,X'X\Pi + \text{plim}\,\frac{1}{T}\,X'\varepsilon(-\Gamma^{-1}) = \underset{\sim}{Q}\Pi$$

Also

$$\text{plim}\,\frac{1}{T}\,Y'\varepsilon = \text{plim}\,\frac{1}{T}\,\Pi'X'\varepsilon + \text{plim}\,\frac{1}{T}\,(-\Gamma^{-1})'\varepsilon'\varepsilon = -(\Gamma^{-1})'\Sigma$$

Finally,

$$\text{plim}\,\frac{1}{T}\,Y'Y = \text{plim}\,\frac{1}{T}\,\Pi'X'X\Pi + \text{plim}\,\frac{1}{T}\,V'X'\Pi + \text{plim}\,\frac{1}{T}\,\Pi'X'V' + \text{plim}\,\frac{1}{T}\,V'V$$

$$= \Pi'\underset{\sim}{Q}\Pi + \Omega$$

6. THEOREM: Let $\hat{\Pi} = (X'X)^{-1}X'Y$ be the OLS estimator of the reduced-form parameter matrix Π. Then $\hat{\Pi}$ is consistent. Furthermore, the asymptotic distribution of $\sqrt{T}\,\text{vec}(\hat{\Pi} - \Pi)$ is $N(0, \Omega \otimes \underset{\sim}{Q}^{-1})$.

Proof: $\hat{\Pi} = \Pi + (X'X)^{-1}X'V$. Hence

$$\text{plim}\,\hat{\Pi} = \Pi + \text{plim}\left(\frac{X'X}{T}\right)^{-1}\frac{X'V}{T} = \Pi + \underset{\sim}{Q}^{-1}\cdot 0 = \Pi$$

Also

$$\sqrt{T}\,\text{vec}(\hat{\Pi} - \Pi) = \begin{bmatrix} \left(\frac{X'X}{T}\right)^{-1} \left(\frac{X'V_{\cdot 1}}{\sqrt{T}}\right) \\ \vdots \\ \left(\frac{X'X}{T}\right)^{-1} \left(\frac{X'V_{\cdot G}}{\sqrt{T}}\right) \end{bmatrix}$$

$$= \left[I_G \otimes \left(\frac{X'X}{T}\right)^{-1}\right] \text{vec}\,\frac{X'V}{\sqrt{T}}$$

Now $I_G \otimes (X'X/T)^{-1}$ converges to $I_G \otimes \underset{\sim}{Q}^{-1}$, while

$$\text{vec}\ \frac{X'V}{\sqrt{T}}$$

converges in distribution to $N(0, \Omega \otimes \underset{\sim}{Q})$. Therefore the asymptotic distribution of $\sqrt{T}\ \text{vec}(\hat{\Pi} - \Pi)$ is normal with zero mean and with covariance matrix

$$(I_G \otimes \underset{\sim}{Q}^{-1})(\Omega \otimes Q)(I_G \otimes \underset{\sim}{Q}^{-1}) = \Omega \otimes [\underset{\sim}{Q}^{-1}\underset{\sim}{Q}\,\underset{\sim}{Q}^{-1}] = \Omega \otimes \underset{\sim}{Q}^{-1}$$

7. NOTATION: Consider a particular structural equation, say the i-th, which is

$$Y\Gamma_{\cdot i} + X\Delta_{\cdot i} + \varepsilon_{\cdot i} = 0$$

Now, some elements in $\Gamma_{\cdot i}$ and $\Delta_{\cdot i}$ will generally be known to be zero; also, we wish to identify one endogenous variable as the "left-hand side" variable by letting its associated coefficient equal minus one. With rearrangement, if necessary, write

$$Y = (y_i, Y_i, Y_i^*)$$

$$\Gamma_{\cdot i} = \begin{pmatrix} -1 \\ \gamma_i \\ 0 \end{pmatrix}$$

$$X = (X_i, X_i^*)$$

$$\Delta_{\cdot i} = \begin{pmatrix} \delta_i \\ 0 \end{pmatrix}$$

Then the i-th equation can be written as

$$y_i = Y_i\gamma_i + X_i\delta_i + \varepsilon_{\cdot i}$$

Sometimes, when we are dealing only with one equation at a time, so that no confusion will result, we will simply write

$$y = Y_i\gamma + X_i\delta + \varepsilon$$

The number of "included" and "excluded" endogenous variables will be denoted by g_* and $g_{**} = G - g_*$, respectively. The number of "included" and "excluded" predetermined variables will be k_* and $k_{**} = K - k_*$. Hence γ_i has dimension $g_* - 1$, and δ_i has dimension k_*.

8. NOTATION: The reduced-form equation $Y = X\Pi + V$ can be correspondingly partitioned as

$$(y_i, Y_i, Y_i^*) = (X_i, X_i^*)\begin{pmatrix} \Pi_{11} & \Pi_{12} & \Pi_{13} \\ \Pi_{21} & \Pi_{22} & \Pi_{23} \end{pmatrix} + (v_i, V_i, V_i^*)$$

(Clearly there is such a partition for each equation.) Also, the usual convention will be used to denote the (partitioned) rows and columns of Π in this partition; for example,

$$\Pi_{\cdot 1} = \begin{pmatrix} \Pi_{11} \\ \Pi_{21} \end{pmatrix} \qquad \Pi_{1\cdot} = [\Pi_{11}, \Pi_{12}, \Pi_{13}]$$

9. THEOREM (Simultaneous Equations Problem): Let the equation $y = Y_i\gamma + X_i\delta + \varepsilon_{\cdot i}$ be written as

$$y = Z\beta + \varepsilon_{\cdot i}$$

where

$$Z = (Y_i, X_i) \qquad \beta = \begin{pmatrix} \gamma \\ \delta \end{pmatrix}$$

Let $\hat{\beta} = (Z'Z)^{-1}Z'y$, the OLS estimator of β. Then

$$\operatorname{plim} \hat{\beta} = \beta + \begin{pmatrix} \Pi_{\cdot 2}' \underset{\sim}{Q} \Pi_{\cdot 2} + \Omega_{**} & \Pi_{\cdot 2}' \underset{\sim}{Q}_i \\ \underset{\sim}{Q}_i' \Pi_{\cdot 2} & \underset{\sim}{Q}_{ii} \end{pmatrix}^{-1} \begin{pmatrix} -(\Gamma_{**}^{-1})'\Sigma_{\cdot i} \\ 0 \end{pmatrix}$$

$$\neq \beta \quad \text{(in general)}$$

where

$$\Pi_{\cdot 2} = \begin{pmatrix} \Pi_{12} \\ \Pi_{22} \end{pmatrix}$$

$$Q = \text{plim}\,\frac{X'X}{T} \qquad \underset{\sim}{Q}_i = \text{plim}\,\frac{X'X_i}{T} \qquad \underset{\sim}{Q}_{ii} = \text{plim}\,\frac{X_i'X_i}{T}$$

$\Sigma_{\cdot i}$ is the i-th column of Σ

Ω_{**} is the $(g_* - 1) \times (g_* - 1)$ submatrix of Ω corresponding to Y_i (or, equivalently, V_i)

Γ_{**}^{-1} is the $G \times (g_* - 1)$ submatrix of Γ^{-1} whose columns correspond to Y_i

Proof:

$$\hat{\beta} = \beta + \begin{pmatrix} Y_i'Y_i & Y_i'X_i \\ X_i'Y_i & X_i'X_i \end{pmatrix}^{-1} \begin{pmatrix} Y_i'\varepsilon_{\cdot i} \\ X_i'\varepsilon_{\cdot i} \end{pmatrix}$$

$$\text{plim}\,\hat{\beta} = \beta + \text{plim}\begin{pmatrix} \frac{Y_i'Y_i}{T} & \frac{Y_i'X_i}{T} \\ \frac{X_i'Y_i}{T} & \frac{X_i'X_i}{T} \end{pmatrix}^{-1} \text{plim}\begin{pmatrix} \frac{Y_i'\varepsilon_{\cdot i}}{T} \\ \frac{X_i'\varepsilon_{\cdot i}}{T} \end{pmatrix}$$

We will now use Proposition 5 to evaluate these probability limits. First consider

$$\text{plim}\,\frac{Y_i'Y_i}{T}$$

From Proposition 3,

$$\text{plim}\,\frac{Y'Y}{T} = \text{plim}\,\frac{(y, Y_i, Y_i^*)'(y, Y_i, Y_i^*)}{T} = \Pi'\underset{\sim}{Q}\Pi + \Omega$$

Hence

$$\text{plim}\,\frac{Y_i'Y_i}{T}$$

is just the submatrix of $\Pi'\underset{\sim}{Q}\Pi + \Omega$ corresponding to Y_i; that is, $\Pi_{\cdot 2}'\underset{\sim}{Q}\Pi_{\cdot 2} + \Omega_{**}$.

Similarly,

$$\text{plim}\ \frac{Y'X}{T} = \Pi'Q$$

from Proposition 3, and

$$\text{plim}\ \frac{Y_i'X_i}{T}$$

is the submatrix of $\Pi'Q$ corresponding to Y_i and X_i, $\Pi_{\cdot 2}'Q_i$. Also

$$\text{plim}\ \frac{X_i'X_i}{T} = Q_{ii}$$

a submatrix of Q.

Finally,

$$\text{plim}\ \frac{Y'\varepsilon}{T} = -(\Gamma^{-1})'\Sigma$$

Therefore

$$\text{plim}\ \frac{Y_i'\varepsilon_{\cdot i}}{T}$$

is the submatrix of $-(\Gamma^{-1})'\Sigma$ corresponding to Y_i and $\varepsilon_{\cdot i}$, namely $-\Gamma_{**}^{-1}\Sigma_{\cdot i}$. And

$$\text{plim}\ \frac{X_i'\varepsilon_{\cdot i}}{T}$$

is zero, completing the proof.

The point of the above theorem is of course that OLS is an *inconsistent* estimator in the context of a system of simultaneous equations.

4.4 IDENTIFICATION

1. DEFINITION: A parameter of a model is *identified* (or identifiable if and only if its value can be deduced from knowledge of the likelihood function of the model. If a parameter is not identified, it is *underidentified*.

2. THEOREM: Under Assumption 1 of Section 4.1, the likelihood function of the model $Y\Gamma + X\Delta + \varepsilon = 0$ is

$$(2\pi)^{-GT/2}|\Sigma|^{-T/2}||\Gamma||^{T} \exp[-\tfrac{1}{2} \operatorname{trace} \Sigma^{-1}(Y\Gamma + X\Delta)'(Y\Gamma + X\Delta)]$$

or equivalently,

$$(2\pi)^{-GT/2}|\Omega|^{-T/2} \exp[-\tfrac{1}{2} \operatorname{trace} \Omega^{-1}(Y - X\Pi)'(Y - X\Pi)]$$

Proof: Consider the t-th observation on the model, $Y_{t.}\Gamma + X_{t.}\Delta + \varepsilon_{t.} = 0$. Since $\varepsilon'_{t.}$ is distributed as $N(0,\Sigma)$, its likelihood function is

$$(2\pi)^{-G/2}|\Sigma|^{-\frac{1}{2}} \exp(-\tfrac{1}{2} \varepsilon_{t.}\Sigma^{-1}\varepsilon'_{t.})$$

Now in order to make the transformation from $\varepsilon_{t.}$ to $Y_{t.}$, note that

$$\frac{\partial\varepsilon_{ti}}{\partial Y_{tj}} = -\Gamma_{ji}$$

in light of the fact that $\varepsilon_{t.} = -Y_{t.}\Gamma - X_{t.}\Delta$. Therefore the Jacobian of the transformation is

$$\left|\left|\frac{\partial\varepsilon_{ti}}{\partial Y_{tj}} \qquad i, j = 1, 2, \cdots, G\right|\right| = ||-\Gamma'|| = ||\Gamma||$$

Therefore the likelihood function associated with the t-th observation can be written

$$(2\pi)^{-G/2}|\Sigma|^{-\frac{1}{2}}||\Gamma|| \exp[-\tfrac{1}{2}(Y_{t.}\Gamma + X_{t.}\Delta)\Sigma^{-1}(Y_{t.}\Gamma + X_{t.}\Delta)']$$

For all T observations on the model, the likelihood function is just the product of these "individual" likelihood functions,

$$(2\pi)^{-GT/2}|\Sigma|^{-T/2}||\Gamma||^{T} \exp[-\tfrac{1}{2} \sum_{t=1}^{T} (Y_{t.}\Gamma + X_{t.}\Delta)\Sigma^{-1}(Y_{t.}\Gamma + X_{t.}\Delta)']$$

$$= (2\pi)^{-GT/2}|\Sigma|^{-T/2}||\Gamma||^{T} \exp[-\tfrac{1}{2} \operatorname{trace}(Y\Gamma + X\Delta)\Sigma^{-1}(Y\Gamma + X\Delta)']$$

$$= (2\pi)^{-GT/2}|\Sigma|^{-T/2}||\Gamma||^{T} \exp[-\tfrac{1}{2} \operatorname{trace} \Sigma^{-1}(Y\Gamma + X\Delta)'(Y\Gamma + X\Delta)]$$

This proves the first part of the theorem.

To show that the second expression in the statement of the Theorem is indeed equivalent to the first, note that

$$|\Omega| = |(\Gamma^{-1})'\Sigma\Gamma^{-1}| = |\Sigma|\ ||\Gamma||^{-2}$$

so that

$$|\Omega|^{-T/2} = |\Sigma|^{-T/2}\ ||\Gamma||^{T}$$

Also

$$\begin{aligned}
&\text{trace } \Omega^{-1}(Y - X\Pi)'(Y - X\Pi) \\
&\quad = \text{trace}(Y - X\Pi)\Omega^{-1}(Y - X\Pi)' \\
&\quad = \text{trace}(Y + X\Delta\Gamma^{-1})(\Gamma\Sigma^{-1}\Gamma')(Y + X\Delta\Gamma^{-1})' \\
&\quad = \text{trace}(Y\Gamma + X\Delta)\Sigma^{-1}(Y\Gamma + X\Delta)' \\
&\quad = \text{trace } \Sigma^{-1}(Y\Gamma + X\Delta)'(Y\Gamma + X\Delta)
\end{aligned}$$

With these substitutions, the two expressions in the statement of the theorem are seen to be equivalent.

3. PROPOSITION: Knowledge of the likelihood function for the model $Y\Gamma + X\Delta + \varepsilon = 0$ is equivalent to knowledge of the reduced-form parameters (Π and Ω).

Proof: It is of course clear that knowledge of Π and Ω imply knowledge of the likelihood function (for any X).

To see the converse, note that $E(Y|X) = X\Pi$ can easily be calculated, for any X, given the likelihood function. But for $T \geq K$, $X\Pi$ uniquely determines Π. Similarly, the elements of Ω are easily calculated, given the likelihood function, since they are just the variances and covariances in the distribution of Y given X.

4. COROLLARY: The reduced-form parameters are identified.

It follows that the reduced-form parameters contain all the information that can conceivably be drawn from any number of observations on the model. This is just a reflection of the fact that all that observations can hope to reveal is the distribution of Y given X.

We therefore have the following definition, which in the present context is equivalent to Definition 1.

5. DEFINITION: A parameter of the structural model $Y\Gamma + X\Delta + \varepsilon$ is identified if and only if it can be deduced from knowledge of the reduced-form parameters Π and Ω.

6. THEOREM: Without a priori restrictions on Γ, Δ, and Σ, none of the parameters of the structural model $Y\Gamma + X\Delta + \varepsilon = 0$ are identified.

Proof: The structural model $Y\Gamma + X\Delta + \varepsilon$ has reduced-form $Y = -X\Delta\Gamma^{-1} - \varepsilon\Gamma^{-1}$. Now choose any nonsingular $G \times G$ matrix A, and consider the structural model $Y\Gamma A + X\Delta A + \varepsilon A = 0$. Its reduced form is

$$\begin{aligned} Y &= -X\Delta AA^{-1}\Gamma^{-1} - \varepsilon AA^{-1}\Gamma^{-1} \\ &= -X\Delta\Gamma^{-1} - \varepsilon\Gamma^{-1} \end{aligned}$$

Hence the models $Y\Gamma + X\Delta + \varepsilon = 0$ and $Y\Gamma A + X\Delta A + \varepsilon A = 0$ have the same reduced form and hence the same likelihood function; they are indistinguishable on the basis of observational information.

We can note that in the proof above, each equation in the system $Y\Gamma A + X\Delta A + \varepsilon A = 0$ is a linear combination of the equations in the original system $Y\Gamma + X\Delta + \varepsilon = 0$. This suggests the following important result.

7. PROPOSITION: Write the model $Y\Gamma + X\Delta + \varepsilon = 0$ as

$$Z\beta + \varepsilon = 0$$

where

$$Z = (Y,X) \qquad \beta = \begin{bmatrix} \Gamma \\ \Delta \end{bmatrix}$$

Let $\beta_{\cdot i}$ be the i-th column of β. Then an arbitrary $(G + K)$-vector α can be distinguished from $\beta_{\cdot i}$ on the basis of observational information alone if and only if α is not a linear combination of columns of β.

Proof: The proof of Theorem 6 has actually proved half of the proposition, since it shows that α is indistinguishable from $\beta_{\cdot i}$ on the basis of observational information if it is a linear combination of the columns of β. What remains to be shown is that if α is indistinguishable from $\beta_{\cdot i}$ on the basis of observational information, then α must be a linear combination of the columns of β.

To show this, recall that all possible observational information is contained in the reduced-form parameters Π and Ω. Furthermore, we know that

$$\Pi\Gamma = -\Delta$$

or

$$(\Pi, I)\beta = 0$$

so that certainly it is true that $(\Pi, I)\beta_{\cdot i} = 0$. Hence if α is to be indistinguishable from $\beta_{\cdot i}$ on the basis of observational information, it must be the case that $(\Pi, I)\alpha = 0$. Therefore α is in the null space of (Π, I). But (Π, I) has $G + K$ columns and has rank K, and so the null space of (Π, I) has rank G [see Note 1]. Finally, β is of rank G and is in the null space of (Π, I), which implies that β spans the null space of (Π, I); that is, any other vector α in the null space of (Π, I) is a linear combination of columns of β. This proves the proposition.

The above proposition is important because it shows that in considering the identification of the parameters of an equation of our structural model, we need consider only *linear* transformations of the model. This motivates the following definition.

8. DEFINITION: A linear transformation of the model $Y\Gamma + X\Delta + \varepsilon = 0$ (or $Z\beta + \varepsilon = 0$), denoted by the $G \times G$ nonsingular transformation matrix A, is ***admissible*** if and only if:

(i) $\beta A = \begin{bmatrix} \Gamma A \\ \Delta A \end{bmatrix}$ satisfies all a priori restrictions on $\beta = \begin{bmatrix} \Gamma \\ \Delta \end{bmatrix}$.

(ii) $A'\Sigma A$ satisfies all a priori restrictions on Σ [see Note 2].

Let us now suppose that we consider one particular equation in the model--for simplicity, say, the first equation. Suppose also that as yet no specific normalization rule has been applied to the equation. Then for the parameters of the first equation to be identified, what is required is that any admissible transformation change $\beta_{\cdot 1}$ by at most multiplying it by a scalar.

9. COROLLARY: The first equation in the model $Z\beta + \varepsilon = 0$ is identified (meaning the parameters of the first equation are all identified) if and only if the first column of any admissible transformation matrix is a vector whose first element is an arbitrary (non-zero) scalar and whose remaining elements all equal zero.

Linear Homogeneous Restrictions

Throughout this subsection, we will be concerned only with the identification of the first equation of the model $Z\beta + \varepsilon = 0$. We will suppose that there are no a priori restrictions on Σ, but that there are R restrictions on $\beta_{\cdot 1}$ which are of the form

$$\phi\beta_{\cdot 1} = 0$$

where ϕ is a *known* $R \times (G + K)$ matrix. Each row of ϕ corresponds to a linear homogeneous restrictions on $\beta_{\cdot 1}$. Such a restriction will be called a *zero restriction* if the corresponding row of ϕ has only one nonzero element.

10. NOTATION:

$$\beta = [\beta_{\cdot 1}, \beta_{(1)}] \qquad \phi\beta = [\phi\beta_{\cdot 1}, \phi\beta_{(1)}]$$

11. LEMMA: $\operatorname{rank}(\phi\beta) = \operatorname{rank}[\phi\beta_{(1)}] \le \min(R, G - 1)$.

Proof: Since $\phi\beta = [\phi\beta_{\cdot 1}, \phi\beta_{(1)}]$ and $\phi\beta_{\cdot 1} = 0$, it is clear that $\operatorname{rank}(\phi\beta) = \operatorname{rank}(\phi\beta_{(1)})$. Also, $\phi\beta_{(1)}$ is of dimension $R \times (G - 1)$, and so its rank cannot exceed R or G - 1.

12. THEOREM (Rank Condition): A necessary and sufficient condition for the first equation to be identified by the a priori restrictions $\phi\beta_{\cdot 1} = 0$ is that

$$\text{rank}(\phi\beta) = G - 1$$

Proof: Let A be a nonsingular transformation matrix with first column A_1; let A_1^* be all of A_1 except the first element, so that

$$A_1 = \begin{bmatrix} c \\ A_1^* \end{bmatrix}$$

where c is a scalar. Then according to Corollary 9 the first equation is identified if and only if $A_1^* = 0$ for any admissible transformation.

Now, the transformation is admissible if and only if $\phi\beta A_1 = 0$. But

$$\phi\beta A_1 = [\phi\beta_{\cdot 1}, \phi\beta_{(1)}]\begin{bmatrix} c \\ A_1^* \end{bmatrix} = \phi\beta_{(1)}A_1^*$$

And a necessary and sufficient condition for $\phi\beta_{(1)}A_1^* = 0$ to imply $A_1^* = 0$ is that $\phi\beta_{(1)}$ have full column rank; that is, $\text{rank}(\phi\beta_{(1)}) = G - 1$. Since $\text{rank}(\phi\beta) = \text{rank}(\phi\beta_{(1)})$, this proves the theorem.

13. COROLLARY (Order Condition): A necessary condition for the first equation to be identified is that $R \geq G - 1$.

14. COMMENT: The order condition is not sufficient for identification There are at least two ways in which the order condition could hold and yet the rank condition could fail:

(i) The coefficients of another equation could satisfy all R restrictions on the first equation. That is, we would have $\phi\beta_{\cdot i} = 0$ for $i \neq 1$. In this case $\text{rank}(\phi\beta) \leq G - 2$ since at least two columns of $\phi\beta$ are zero; the first equation cannot be identified.

(ii) All G - 1 other equations may satisfy a particular restriction on the first equation. Then a row of $\phi\beta$ would equal zero; such a restriction does not aid in the identification of the first equation since it does not add to the rank of $\phi\beta$.

15. COMMENT: Note that if all R restrictions are zero restrictions ("exclusion restrictions"), then $\phi\beta_{(1)}$ is just composed of the coefficients in the other equations of the variables that are excluded from the first equation.

16. DEFINITION: The first equation is *underidentified* if $\text{rank}(\phi\beta) < G - 1$. It is *exactly identified* if $\text{rank}(\phi\beta) = G - 1$ and $\text{rank}(\phi) = G - 1$, and it is *overidentified* if $\text{rank}(\phi\beta) = G - 1$ and $\text{rank}(\phi) > G - 1$.

The above results on identification under linear homogeneous restrictions can also be derived in another (equivalent) manner.

17. THEOREM: A necessary and sufficient condition for the identification of the first equation under the restrictions $\phi\beta_{\cdot 1} = 0$ is

$$\text{rank}\begin{bmatrix} \Pi, I_K \\ \phi \end{bmatrix} = G + K - 1$$

Proof: Again, all possible observational information is contained in the reduced-form parameters. Since we are assuming that there is no a priori knowledge of Σ, all that is available is the knowledge that $\Pi\Gamma = -\Delta$ [or $(\Pi, I)\beta = 0$] and that $\phi\beta_{\cdot 1} = 0$. That is, the available information--both a priori and observational--on $\beta_{\cdot 1}$ is only that

$$\begin{bmatrix} \Pi, I \\ \phi \end{bmatrix}\beta_{\cdot 1} = 0$$

Hence the first equation is identified if and only if this equation can be solved for $\beta_{\cdot 1}$, and if the solution is unique up to scalar multiplication. But since $\beta_{\cdot 1}$ is of dimension $G + K$, this can be done if and only if

$$\text{rank}\begin{bmatrix} \Pi, I \\ \phi \end{bmatrix} = G + K - 1$$

Note that since the matrix

$$\begin{bmatrix} \Pi, I \\ \phi \end{bmatrix}$$

is of dimension $(R + K) \times (G + K)$, a necessary condition for identification is that $R \geq G - 1$, as before.

The correspondence between the approaches of Theorem 12 and Theorem 17 is contained in the following result.

18. PROPOSITION:

$$\text{rank}\begin{bmatrix} \Pi, I \\ \phi \end{bmatrix} = \text{rank}(\phi\beta) + K$$

Proof: We will use the fact that the rank of a matrix is not changed when the matrix is multiplied by a nonsingular matrix.

$$\text{rank}\begin{bmatrix} \Pi, I_K \\ \phi \end{bmatrix} = \text{rank}\left\{\begin{bmatrix} \Pi, I_K \\ \phi \end{bmatrix}\begin{bmatrix} \Gamma & 0 \\ \Delta & I_K \end{bmatrix}\right\}$$

$$= \text{rank}\begin{bmatrix} 0 & I_K \\ \phi\beta & N \end{bmatrix}$$ (where N is made up of the last K columns of ϕ)

$$= \text{rank}\left\{\begin{bmatrix} I_K & 0 \\ -N & I_R \end{bmatrix}\begin{bmatrix} 0 & I_K \\ \phi\beta & N \end{bmatrix}\right\}$$

$$= \text{rank}\begin{bmatrix} 0 & I_K \\ \phi\beta & 0 \end{bmatrix}$$

$$= \text{rank}(\phi\beta) + K$$

The expression in Theorem 17 can also be "simplified" somewhat if all the restrictions are zero (exclusion) restrictions.

19. PROPOSITION: Let the restrictions on the first equation be embodied in the partitioning

$$\Gamma_{\cdot 1} = \begin{bmatrix} -1 \\ \gamma \\ 0 \end{bmatrix} \qquad \Delta_{\cdot 1} = \begin{bmatrix} \delta \\ 0 \end{bmatrix}$$

Let Π be partitioned as

$$\Pi = \begin{bmatrix} \Pi_{11} & \Pi_{12} & \Pi_{13} \\ \Pi_{21} & \Pi_{22} & \Pi_{23} \end{bmatrix}$$

as in Notation 8 of Section 4.3. Then a necessary and sufficient condition for the identification of the first equation is that

$$\text{rank}(\Pi_{22}) = g_* - 1$$

(where g_* is the number of endogenous variables appearing in the equation).

Proof: Since $\Pi\Gamma = -\Delta$, we can note that the available a priori and observational information is that

$$\begin{bmatrix} \Pi_{11} & \Pi_{12} & \Pi_{13} \\ \Pi_{21} & \Pi_{22} & \Pi_{23} \end{bmatrix} \begin{bmatrix} -1 \\ \gamma \\ 0 \end{bmatrix} = \begin{bmatrix} -\delta \\ 0 \end{bmatrix}$$

which amounts to

$$\delta = \Pi_{11} - \Pi_{12}\gamma$$

$$\Pi_{22}\gamma = \Pi_{21}$$

Since a normalization rule has now been applied, we require for identification that γ and δ be uniquely determined. Clearly determining δ is no problem, if γ can be determined, and so the problem is whether the set of equations

$$\Pi_{22}\gamma = \Pi_{21}$$

can be solved uniquely for γ. Since the dimension of γ is $g_* - 1$, a necessary and sufficient condition for a unique solution is that $\text{rank}(\Pi_{22}) = g_* - 1$.

Covariance Restrictions

We will again consider the identification of the first equation of the model $Z\beta + \varepsilon = 0$ under the restrictions $\phi\beta_{\cdot 1} = 0$, with the further restrictions that $\sigma_{1j} = 0$ for some $j \neq 1$. That is, the further a priori information here is that the disturbance in the first equation is uncorrelated with the disturbance in some other equation or equations.

20. NOTATION: Let

$$J_2 = \{j \mid \sigma_{1j} = 0 \qquad j = 1, 2, \cdots, G\}$$

$$J_1 = \{j \mid j \notin J_2 \qquad j = 1, 2, \cdots, G\}$$

Equation j is said to be in J_1 if $j \in J_1$, and similarly for J_2. Let the equations in J_1 be the first g_1 equations; the last $g_2 = G - g_1$ equations are then in J_2. Partition Σ and any transformation matrix A as

$$\Sigma = \begin{bmatrix} \Sigma_{11} & \Sigma_{12} \\ \Sigma_{21} & \Sigma_{22} \end{bmatrix} \qquad A = \begin{bmatrix} A_{11} & A_{12} \\ A_{21} & A_{22} \end{bmatrix}$$

where Σ_{11} and A_{11} are $g_1 \times g_1$, and the other submatrices are conformable.

21. COMMENT: Note that the first row of Σ_{12} and the first column of Σ_{21} are zero. Also note that $A'\Sigma A$ satisfies these a priori restrictions on Σ if and only if

$$[A_{12}', A_{22}']\Sigma A_1 = 0$$

where A_1 is again the first column of A.

22. LEMMA: A necessary condition for the identification of the first equation is that for every admissible A,

$$\text{rank}\begin{bmatrix} \phi\beta \\ (A'_{12},A'_{22})\Sigma \end{bmatrix} = G - 1$$

Proof: The a priori restrictions available are that

$$\begin{bmatrix} \phi\beta \\ (A'_{12},A'_{22})\Sigma \end{bmatrix} A_1 = 0$$

For the first equation to be identified we require the solution (A_1) of this set of equations be of the form

$$A_1 = \begin{bmatrix} c \\ 0 \end{bmatrix} \qquad \text{(c a scalar)}$$

A necessary condition for this is that the solution be unique up to scalar multiplication. But this will be the case only if

$$\text{rank}\begin{bmatrix} \phi\beta \\ (A'_{12},A'_{22})\Sigma \end{bmatrix} = G - 1$$

(Note that in the proof, A_{12} and A_{22} are treated as fixed, while A_1 is variable. This is legitimate since A_1 is not part of A_{12} or A_{22}.)

23. THEOREM (Generalized Rank Condition): A necessary (*not* sufficient) condition for the first equation to be identified is that

$$\text{rank}\begin{bmatrix} \phi\beta \\ \Sigma_{21},\Sigma_{22} \end{bmatrix} = G - 1$$

Proof: Since the identity transformation is clearly admissible, we can pick $A = I$. Then

$$[A'_{12}, A'_{22}]\Sigma = [0,I]\Sigma = [\Sigma_{21},\Sigma_{22}]$$

and the theorem follows from Lemma 22.

24. COMMENT: Note that although

$$\begin{bmatrix} \phi\beta \\ \Sigma_{21},\Sigma_{22} \end{bmatrix}$$

has G columns, its maximum rank is G - 1 since its first column is zero. Also note that if some equation j in J_2 is an identity, then a row of $[\Sigma_{21},\Sigma_{22}]$ is zero. Thus there is no aid to the identification of the first equation in pointing out that the disturbance of the first equation is uncorrelated with the (identically zero) disturbance of an identity.

25. COROLLARY (Generalized Order Condition): A necessary condition for the identification of the first equation is

$$R + g_2 \geq G - 1$$

where again R is the number of coefficient restrictions and g_2 is the number of covariance restrictions.

26. DEFINITION: The i-th equation of the model $Z\beta + \varepsilon = 0$ is *identified with respect to the j-th equation* if every admissible transformation A has $a_{ji} = 0$ (where a_{ji} is the j,i-th element of A).

27. COMMENT: What this requires is that any linear combination of equations which is observationally equivalent to the i-th equation does not contain the j-th equation. Also, note that any equation is identified if and only if it is identified with respect to all G - 1 other equations.

28. THEOREM: A *sufficient* condition for the identification of the first equation is that

(i) The generalized rank condition holds.

(ii) Every equation in J_2 is identified with respect to every equation in J_1.

Proof: Again letting

$$A = \begin{bmatrix} A_{11} & A_{12} \\ A_{21} & A_{22} \end{bmatrix}$$

requirement (ii) implies that $A_{12} = 0$. Since for A to be admissible we must have

$$[A'_{12}, A'_{22}] \begin{bmatrix} \Sigma_{11} & \Sigma_{12} \\ \Sigma_{21} & \Sigma_{22} \end{bmatrix} A_1 = 0$$

in this case what is required is that

$$[A'_{22}\Sigma_{21}, A'_{22}\Sigma_{22}]A_1 = 0$$

Since A_{22} is nonsingular (since A is), the requirement is just that

$$[\Sigma_{21}, \Sigma_{22}]A_1 = 0$$

Also required for admissibility is that $\phi\beta A_1 = 0$. Hence the a priori restrictions on A_1 for admissibility are that

$$\begin{bmatrix} \phi\beta \\ \Sigma_{21}, \Sigma_{22} \end{bmatrix} A_1 = 0$$

But

$$A_1 = \begin{bmatrix} c \\ 0 \end{bmatrix} \qquad \text{(c a scalar)}$$

is a solution, and it is unique up to scalar multiplication if and only if the generalized rank condition holds. Hence this is sufficient for the identification of the first equation.

29. COMMENT: We have actually proved a little more than was claimed. What we have proved is that if every equation in J_2 is identified with respect to every equation in J_1, then a necessary and sufficient condition for the identification of the first equation is that the generalized rank condition holds.

Variance Restrictions

Occasionally some disturbance in the model $Z\beta + \varepsilon = 0$ may be identically zero. This is most often the case when the equation is an identity, that is, not only is the disturbance zero, but also all the coefficients in the equation are known a priori. In such a case the equation is clearly identified. The case to be considered here, however, is the one in which the disturbance in the first equation is identically zero, but $\beta_{\cdot 1}$ is not completely known a priori.

30. THEOREM: A necessary and sufficient condition for the identification of the first equation under the restrictions $\phi\beta_{\cdot 1} = 0$ and $\sigma_{11} = 0$ (σ_{11} being the variance of the disturbance of the first equation) is that

$$\text{rank}\begin{bmatrix} \phi\beta \\ \Sigma \end{bmatrix} = G - 1$$

Proof: For the admissibility of the transformation A we require only that

$$\phi\beta A_1 = 0$$

$$A_1'\Sigma A_1 = 0$$

Since Σ is positive semidefinite, this last requirement holds if and only if $\Sigma A_1 = 0$. Hence we require

$$\begin{bmatrix} \phi\beta \\ \Sigma \end{bmatrix} A_1 = 0$$

But

$$A_1 = \begin{bmatrix} c \\ 0 \end{bmatrix}$$

is a solution, and it is unique up to scalar multiplication if and only if

$$\text{rank}\begin{bmatrix} \phi\beta \\ \Sigma \end{bmatrix} = G - 1$$

31. COMMENT: Note that identification is guaranteed in this case as long as $\text{rank}(\Sigma) = G - 1$, which will be the case if the other $G - 1$ disturbances are not linearly dependent.

Identification of the Complete System

Occasionally it is useful to consider identification of all equations at once, rather than of each equation separately. This is the case, for example, if there are cross-equation restrictions, e.g., a restriction that two coefficients in different equations are equal. It is worth noting at the outset that such cross-equation constraints can make sense only after each equation has been normalized. We therefore assume that this is so; specifically, that $\Gamma_{ii} = -1$, $(i = 1, 2, \cdots, G)$.

We will consider linear restrictions on all coefficients in the system. That is, we consider $Y\Gamma + X\Delta + \varepsilon = 0$, or $Z\beta + \varepsilon = 0$, as before, with the restrictions that

$$\phi \text{ vec } \beta = c$$

where ϕ is a known $R \times (G^2 + GK)$ matrix and c is a known vector of dimension R.

32. PROPOSITION: As above, consider the model $Z\beta + \varepsilon = 0$, with prior information that $\phi \text{ vec } \beta = c$. Let N be the $G \times (G^2 + GK)$ matrix defined as follows:

$$N_{ij} = \begin{cases} 1 & \text{if } j = i + (G + K)(i - 1) \\ 0 & \text{otherwise} \end{cases}$$

Then a necessary and sufficient condition for β to be identified under this prior information is that

$$\text{rank}\left\{\begin{bmatrix} \phi \\ N \end{bmatrix}(I_G \otimes \beta)\right\} = G^2$$

Proof: Consider the $G \times G$ transformation matrix A, and transformed model $Z\beta A + \varepsilon A = 0$. For this to be observationally equivalent to the original model we require

$$\phi \text{ vec}(\beta A) = c$$

or, equivalently,

$$\phi(I \otimes \beta)\text{vec } A = c$$

Since all equations have been normalized, we also know that $\Gamma_{ii} = -1$, which means that

$$(\text{vec } \beta)_{i+(G+K)(i-1)} = -1 \qquad (i = 1, 2, \cdots, G)$$

This can be expressed as

$$N \text{ vec } \beta = -e$$

where N is as defined above, and where e is a vector of ones. Therefore, for the transformation A to be admissible, we also require

$$N \text{ vec}(\beta A) = -e$$

or

$$N(I \otimes \beta)\text{vec } A = -e$$

We therefore have, for A to be admissible, the requirements that

$$\phi(I \otimes \beta)\text{vec } A = c$$

and

$$N(I \otimes \beta)\text{vec } A = -e$$

That is,

$$\begin{bmatrix} \phi \\ N \end{bmatrix}(I \otimes \beta)\text{vec } A = \begin{bmatrix} c \\ -e \end{bmatrix}$$

Clearly vec I_G is a solution. It is the unique solution if and only if

$$\text{rank}\left\{\begin{bmatrix} \phi \\ N \end{bmatrix}(I \otimes \beta)\right\} = G^2$$

(Note that we require a unique solution, rather than simply a solution which is unique up to scalar multiplication, since the normalization has already been applied.)

4.5 INDIRECT LEAST SQUARES

Let us now turn to the problem of the estimation of the coefficients of a single equation, say the first equation. That is, we wish to estimate the equation

$$Y\Gamma_{\cdot 1} + X\Delta_{\cdot 1} + \varepsilon_{\cdot 1} = 0$$

(or $Z\beta_{\cdot 1} + \varepsilon_{\cdot 1} = 0$). We will suppose that the a priori (identifying) information available is of the form $\phi\beta_{\cdot 1} = 0$.

1. DEFINITION: The *indirect least squares* (ILS) estimator of $\beta_{\cdot 1}$ is the solution to

$$\begin{bmatrix} \hat{\Pi}, I \\ \phi \end{bmatrix} \beta_{\cdot 1} = 0$$

where $\hat{\Pi} = (X'X)^{-1}X'Y$, if the equation is identified. If the equation is overidentified, it is understood that certain rows of $(\hat{\Pi}, I)$ will be dropped (ignored) to obtain the solution.

2. PROPOSITION: The ILS estimator is consistent. It is also unique if the equation is exactly identified.

Proof: Consider first the case in which the equation is exactly identified. Hence $R = G - 1$; the matrix

$$W = \begin{bmatrix} \hat{\Pi}, I \\ \phi \end{bmatrix}$$

is of dimension $(G + K - 1) \times (G + K)$ and with probability one its rank is $G + K - 1$. Hence the solution which defines the ILS estimator is unique up to scalar multiplication; as soon as a particular normalization is applied, the ILS estimator is unique. This proves the second part of the proposition.

To prove consistency, partition $\hat{W}$ as

$$\hat{W} = [\hat{W}_{\cdot 1}, \hat{W}_{(1)}]$$

Let the first element of $\beta_{\cdot 1}$ equal minus one, and let

$$\beta_{\cdot 1} = \begin{bmatrix} -1 \\ \beta^*_{\cdot 1} \end{bmatrix}$$

Then clearly the equation $\hat{W}\beta_{\cdot 1} = 0$ is equivalent to $\hat{W}_{(1)}\beta^*_{\cdot 1} = \hat{W}_{\cdot 1}$. The solution is

$$\hat{\beta}^*_{\cdot 1} = \hat{W}^{-1}_{(1)}\hat{W}_{\cdot 1}$$

the necessary inverse existing with probability one. Hence we have

$$\text{plim } \hat{\beta}^*_{\cdot 1} = W^{-1}_{(1)}W_{\cdot 1}$$

where

$$W = [W_{\cdot 1}, W_{(1)}] = \text{plim } \hat{W} = \begin{bmatrix} \Pi, I \\ \phi \end{bmatrix}$$

since $\hat{\Pi}$ is a consistent estimator of Π. But since $W\beta_{\cdot 1} = 0$, it follows that

$$\beta^*_{\cdot 1} = W^{-1}_{(1)}W_{\cdot 1} = \text{plim } \hat{\beta}^*_{\cdot 1}$$

which proves consistency.

In the overidentified case, $R > G - 1$; so the row dimension of $\hat{W}$ is at least $G + K$. Hence we will generally have $\text{rank}(\hat{W}) = G + K$ even though $\text{rank}(W) = G + K - 1$. In this case *no* solution to the equation $\hat{W}\beta_{\cdot 1} = 0$ exists. However, if we drop (any) $R - G + 1$ rows of $(\hat{\Pi}, I)$, call the remaining matrix $(\hat{\Pi}, I)^o$, and define

$$\hat{W}^o = \begin{bmatrix} (\hat{\Pi}, I)^o \\ \phi \end{bmatrix}$$

then $\hat{W}^o$ has $G + K - 1$ rows and rank $G + K - 1$ (with probability one). Then the equation $\hat{W}^o\beta_{\cdot 1} = 0$ has a unique solution, and this solution is a consistent estimate of $\beta_{\cdot 1}$ by the previous line of proof.

Note that if the equation were underidentified, $\hat{W}\beta_{\cdot 1} = 0$ would have an infinite number of nonequivalent solutions--none of which would be a consistent estimator of $\beta_{\cdot 1}$.

When the only restrictions on $\beta_{\cdot 1}$ are exclusion restrictions, things can be "simplified" somewhat. Again letting

$$\Gamma_{\cdot 1} = \begin{pmatrix} -1 \\ \gamma \\ 0 \end{pmatrix} \qquad \Delta_{\cdot 1} = \begin{pmatrix} \delta \\ 0 \end{pmatrix} \qquad \Pi = \begin{pmatrix} \Pi_{11} & \Pi_{12} & \Pi_{13} \\ \Pi_{21} & \Pi_{22} & \Pi_{23} \end{pmatrix}$$

then ILS amounts to solving the following equations for γ and δ:

$$\begin{pmatrix} \hat{\Pi}_{11} & \hat{\Pi}_{12} & \hat{\Pi}_{13} \\ \hat{\Pi}_{21} & \hat{\Pi}_{22} & \hat{\Pi}_{23} \end{pmatrix} \begin{pmatrix} -1 \\ \gamma \\ 0 \end{pmatrix} = \begin{pmatrix} -\delta \\ 0 \end{pmatrix}$$

or

$$\delta = \hat{\Pi}_{11} - \hat{\Pi}_{12}\gamma$$

$$\hat{\Pi}_{22}\gamma = \hat{\Pi}_{21}$$

Clearly the problem is solving $\hat{\Pi}_{22}\gamma = \hat{\Pi}_{21}$ for γ. Note that Π_{22} has dimension $k_{**} \times g_* - 1$ (where k_{**} is the number of excluded predetermined variables and g_* is the number of included endogenous variables). If the equation is exactly identified, $\hat{\Pi}_{22}$ is square and nonsingular (with probability one), so that

$$\hat{\gamma} = \hat{\Pi}_{22}^{-1}\hat{\Pi}_{21}$$

is unique and has probability limit

$$\text{plim } \hat{\gamma} = \Pi_{22}^{-1}\Pi_{21} = \gamma$$

Also $\text{plim } \hat{\delta} = \Pi_{11} - \Pi_{12}\gamma = \delta$. On the other hand, if the equation is overidentified, no solution exists unless some rows of Π_{22} are dropped.

3. COMMENT: Although ILS is consistent, it is apparent that it is not a very appropriate estimator in the overidentified case. When the equation is overidentified, ILS ignores some of the sample information and is therefore clearly inefficient. This is quite apart from the other consideration that the estimator is not uniquely defined, that is, that the choice of which sample information to ignore is arbitrary.

Finally, we now show that the indirect least squares estimator is an instrumental variables estimator. We will see later (Proposition 20 of Section 4.6) that this demonstrates the inefficiency of the indirect least squares estimator in the overidentified case.

4. PROPOSITION: The indirect least squares estimator is an instrumental variables estimator.

Proof: Indirect least squares is the solution to

$$\hat{\Pi}\begin{pmatrix} 1 \\ \hat{\gamma} \\ 0 \end{pmatrix} = \begin{pmatrix} -\hat{\delta} \\ 0 \end{pmatrix}$$

where $\hat{\Pi} = (X'X)^{-1}X'Y$. Since $Y = (y, Y_1, Y_1^*)$, we can write this as

$$(X'X)^{-1}X'(y, Y_1)\begin{pmatrix} -1 \\ \hat{\gamma} \end{pmatrix} = \begin{pmatrix} -\hat{\delta} \\ 0 \end{pmatrix}$$

or

$$(X'X)^{-1}X'Y_1\hat{\gamma} + \begin{pmatrix} \hat{\delta} \\ 0 \end{pmatrix} = (X'X)^{-1}X'y$$

In the overidentified case there are too many equations, as we have seen, and so we need to drop $k_{**} - g_* + 1$ equations. To do so, let S be a $(k_* + g_* - 1) \times K$ "selection" matrix; that is, each row of S contains all zeroes except for a one in the column corresponding to the position of one of the $k_* + g_* - 1$ equations *not* being dropped. (In the exactly identified case, we have $S = I_K$.) We then wish to solve the reduced system

$$S(X'X)^{-1}X'Y_1\hat{\gamma} + S\begin{pmatrix}\hat{\delta}\\0\end{pmatrix} = S(X'X)^{-1}X'y$$

Note that

$$S\begin{pmatrix}\hat{\delta}\\0\end{pmatrix} = S(X'X)^{-1}X'X\begin{pmatrix}\hat{\delta}\\0\end{pmatrix}$$

$$= S(X'X)^{-1}X'X_1\hat{\delta}$$

Hence we have

$$S(X'X)^{-1}X'Y_1\hat{\gamma} + S(X'X)^{-1}X'X_1\hat{\delta} = S(X'X)^{-1}X'y$$

The solution is

$$\begin{pmatrix}\hat{\gamma}\\\hat{\delta}\end{pmatrix} = \{[S(X'X)^{-1}X'](Y_1,X_1)\}^{-1}[S(X'X)^{-1}X']y$$

This is an instrumental variables estimator, where the set of instruments for (Y_1,X_1) is $X(X'X)^{-1}S'$.

4.6 TWO-STAGE LEAST SQUARES

Let us consider the estimation of a single equation, say the first, in a set of simultaneous equations. In the notation of Notation 7 of Section 4.3, the equation to be estimated is

$$y_1 = Y_1\gamma_1 + X_1\delta_1 + \varepsilon_{\cdot 1}$$

or else

$$y = Y_1\gamma + X_1\delta + \varepsilon_{\cdot 1}$$

1. DEFINITION: The *two-stage least squares* (2SLS) estimator of γ and δ is the instrumental variables estimator with X_1 serving as its own instrument and $\hat{Y}_1 = X(X'X)^{-1}X'Y_1$ serving as instrument for Y_1. Explicitly,

$$\begin{pmatrix} \hat{\gamma} \\ \hat{\delta} \end{pmatrix}_{2SLS} = [(\hat{Y}_1, X_1)'(Y_1, X_1)]^{-1}(\hat{Y}_1, X_1)'y$$

$$= \begin{pmatrix} \hat{Y}_1'Y_1 & \hat{Y}_1'X_1 \\ X_1'Y_1 & X_1'X_1 \end{pmatrix}^{-1} \begin{pmatrix} \hat{Y}_1'y \\ X_1'y \end{pmatrix}$$

2. LEMMA: $\hat{Y}_1'X_1 = Y_1'X_1$.

Proof: Since $\hat{Y}_1 = X(X'X)^{-1}X'Y_1$, $\hat{Y}_1'X = Y_1'X$. Hence $\hat{Y}_1'X_1 = Y_1'X_1$.

3. COROLLARY: The 2SLS estimator of γ and δ can be written

$$\begin{pmatrix} \hat{\gamma} \\ \hat{\delta} \end{pmatrix}_{2SLS} = \begin{pmatrix} Y_1'X(X'X)^{-1}X'Y_1 & Y_1'X_1 \\ X_1'Y_1 & X_1'X_1 \end{pmatrix}^{-1} \begin{pmatrix} Y_1'X(X'X)^{-1}X'y \\ X_1'y \end{pmatrix}$$

4. COROLLARY:

$$\begin{pmatrix} \hat{\gamma} \\ \hat{\delta} \end{pmatrix}_{2SLS} = \begin{pmatrix} \gamma \\ \delta \end{pmatrix} + \begin{pmatrix} Y_1'X(X'X)^{-1}X'Y_1 & Y_1'X_1 \\ X_1'Y_1 & X_1'X_1 \end{pmatrix}^{-1} \begin{pmatrix} Y_1'X(X'X)^{-1}X'\varepsilon_{\cdot 1} \\ X_1'\varepsilon_{\cdot 1} \end{pmatrix}$$

5. LEMMA:

$$\operatorname{plim} \frac{1}{T} \begin{pmatrix} Y_1'X(X'X)^{-1}X'Y_1 & Y_1'X_1 \\ X_1'Y_1 & X_1'X_1 \end{pmatrix} = \begin{pmatrix} \Pi_{\cdot 2}'\underline{Q}\Pi_{\cdot 2} & \Pi_{\cdot 2}'\underline{Q}_1 \\ \underline{Q}_1'\Pi_{\cdot 2} & \underline{Q}_{11} \end{pmatrix}$$

Furthermore, if the equation is identified (by the exclusion restrictions already imposed), this matrix is nonsingular.

Proof: From Proposition 5 of Section 4.3,

$$\operatorname{plim} \frac{Y'X}{T} = \Pi'\underline{Q}$$

Hence

$$\operatorname{plim} \frac{Y_1'X}{T} = \Pi_{\cdot 2}'\underline{Q}$$

$$\operatorname{plim} \frac{Y_1'X_1}{T} = \Pi_{\cdot 2}'\underline{Q}_1$$

where

$$\text{plim}\ \frac{X'X}{T} = \underset{\sim}{Q}$$

$$\text{plim}\ \frac{X'X_1}{T} = \underset{\sim}{Q}_1$$

$$\text{plim}\ \frac{X_1'X_1}{T} = \underset{\sim}{Q}_{11}$$

The first part of the lemma then follows.

To show the second part, note that since $\underset{\sim}{Q}$ is positive definite, there exists a nonsingular matrix V such that $\underset{\sim}{Q} = V'V$. Then

$$\begin{pmatrix} \Pi_{\cdot 2}'\underset{\sim}{Q}\Pi_{\cdot 2} & \Pi_{\cdot 2}'\underset{\sim}{Q}_1 \\ \underset{\sim}{Q}_1'\Pi_{\cdot 2} & \underset{\sim}{Q}_{11} \end{pmatrix} = \begin{pmatrix} \Pi_{\cdot 2}'V'V\Pi_{\cdot 2} & \Pi_{\cdot 2}'V'V\begin{pmatrix} I \\ 0 \end{pmatrix} \\ [I, 0]V'V\Pi_{\cdot 2} & [I, 0]V'V\begin{pmatrix} I \\ 0 \end{pmatrix} \end{pmatrix}$$

$$= \left(V\begin{pmatrix} \Pi_{\cdot 2}, & \begin{matrix} I \\ 0 \end{matrix} \end{pmatrix} \right)' \left(V\begin{pmatrix} \Pi_{\cdot 2}, & \begin{matrix} I \\ 0 \end{matrix} \end{pmatrix} \right)$$

This is nonsingular if and only if

$$\text{rank}\left(V\begin{pmatrix} \Pi_{\cdot 2}, & \begin{matrix} I \\ 0 \end{matrix} \end{pmatrix} \right) = g_* + k_* - 1$$

(its column dimension). But since V is nonsingular,

$$\text{rank}\left(V\begin{pmatrix} \Pi_{\cdot 2}, & \begin{matrix} I \\ 0 \end{matrix} \end{pmatrix} \right) = \text{rank}\begin{pmatrix} \Pi_{12} & I \\ \Pi_{22} & 0 \end{pmatrix}$$

$$= \text{rank}(\Pi_{22}) + k_*$$

Hence the rank of the matrix above is indeed $g_* + k_* - 1$ if $\text{rank}(\Pi_{22}) = g_* - 1$, that is, if the equation is identified by the exclusion restrictions.

6. THEOREM: If the equation is identified, the 2SLS estimator of γ and δ is consistent.

Proof: Using Corollary 4,

$$\text{plim}\begin{pmatrix} \hat{\gamma} \\ \hat{\delta} \end{pmatrix}_{2SLS} = \begin{pmatrix} \gamma \\ \delta \end{pmatrix} + \left\{\text{plim}\,\frac{1}{T}\begin{pmatrix} Y_1'X(X'X)^{-1}X'Y_1 & Y_1'X_1 \\ X_1'Y_1 & X_1'X_1 \end{pmatrix}\right\}^{-1} \cdot$$

$$\left\{\text{plim}\,\frac{1}{T}\begin{pmatrix} Y_1'X(X'X)^{-1}X'\varepsilon_{\cdot 1} \\ X_1'\varepsilon_{\cdot 1} \end{pmatrix}\right\}$$

Now, the matrix in brackets is finite and nonsingular by Lemma 5, so that all that needs to be shown is that

$$\text{plim}\,\frac{1}{T}\begin{pmatrix} Y_1'X(X'X)^{-1}X'\varepsilon_{\cdot 1} \\ X_1'\varepsilon_{\cdot 1} \end{pmatrix} = 0$$

But

$$\text{plim}\,\frac{1}{T}\,X_1'\varepsilon_{\cdot 1} = 0$$

$$\text{plim}\,\frac{1}{T}\,Y_1'X(X'X)^{-1}X'\varepsilon_{\cdot 1} = \Pi_{\cdot 2}'\underset{\sim}{Q}\,\underset{\sim}{Q}^{-1} \cdot 0 = 0$$

proving the theorem.

7. PROPOSITION: Let σ^2 be the variance of the first equation's disturbance (that is, $\sigma^2 = \sigma_{11}$, the 1,1-th element of Σ), and let

$$s^2 = \frac{1}{T}\,(y - Y_1\hat{\gamma} - X_1\hat{\delta})'(y - Y_1\hat{\gamma} - X_1\hat{\delta})$$

Then s^2 is a consistent estimator of σ^2 if the equation is identified.

Proof:

$$\text{plim}\, s^2 = \text{plim}\,\frac{1}{T}[\varepsilon_{\cdot 1} + Y_1(\gamma - \hat{\gamma}) + X_1(\delta - \hat{\delta})]'[\varepsilon_{\cdot 1} + Y_1(\gamma - \hat{\gamma}) + X_1(\delta - \hat{\delta})]$$

$$= \text{plim}\,\frac{1}{T}\,\varepsilon_{\cdot 1}'\varepsilon_{\cdot 1} + \text{plim}(\gamma - \hat{\gamma})'\,\frac{Y_1'Y_1}{T}\,(\gamma - \hat{\gamma})$$

$$+ \text{plim}(\delta - \hat{\delta})'\,\frac{X_1'X_1}{T}\,(\delta - \hat{\delta}) +$$

$$+ 2\,\text{plim}(\gamma - \hat{\gamma})' \frac{Y_1'X_1}{T}(\delta - \hat{\delta}) + 2\,\text{plim}(\gamma - \hat{\gamma})' \frac{Y_1'\varepsilon_{\cdot 1}}{T}$$

$$+ 2\,\text{plim}(\delta - \hat{\delta})' \frac{X_1'\varepsilon_{\cdot 1}}{T}$$

$$= \sigma^2$$

since $\text{plim}\,\frac{1}{T}\varepsilon_{\cdot 1}'\varepsilon_{\cdot 1} = \sigma^2$, $\text{plim}(\gamma - \hat{\gamma}) = 0$, and $\text{plim}(\delta - \hat{\delta}) = 0$.

8. THEOREM: The asymptotic distribution of

$$\sqrt{T}\begin{pmatrix} \hat{\gamma} - \gamma \\ \hat{\delta} - \delta \end{pmatrix}$$

is $N(0, \sigma^2\overline{B}^{-1})$, where

$$\overline{B} = \begin{pmatrix} \Pi_{\cdot 2}'\underset{\sim}{Q}\Pi_{\cdot 2} & \Pi_{\cdot 2}'\underset{\sim}{Q}_1 \\ \underset{\sim}{Q}_1'\Pi_{\cdot 2} & \underset{\sim}{Q}_{11} \end{pmatrix} = \text{plim}\,\frac{1}{T}\begin{pmatrix} \hat{Y}_1'Y_1 & Y_1'X_1 \\ X_1'Y_1 & X_1'X_1 \end{pmatrix}$$

Proof: Clearly we have

$$\sqrt{T}\begin{pmatrix} \hat{\gamma} - \gamma \\ \hat{\delta} - \delta \end{pmatrix} = \begin{pmatrix} \frac{\hat{Y}_1'Y_1}{T} & \frac{Y_1'X_1}{T} \\ \frac{X_1'Y_1}{T} & \frac{X_1'X_1}{T} \end{pmatrix}^{-1} \begin{pmatrix} \frac{\hat{Y}_1'\varepsilon_{\cdot 1}}{\sqrt{T}} \\ \frac{X_1'\varepsilon_{\cdot 1}}{\sqrt{T}} \end{pmatrix}$$

The first matrix on the right-hand side converges to $\overline{B}^{-1}$. We therefore consider the vector

$$\begin{pmatrix} \frac{\hat{Y}_1'\varepsilon_{\cdot 1}}{\sqrt{T}} \\ \frac{X_1'\varepsilon_{\cdot 1}}{\sqrt{T}} \end{pmatrix} = \begin{pmatrix} \frac{Y_1'X}{T}\left(\frac{X'X}{T}\right)^{-1}\frac{X'\varepsilon_{\cdot 1}}{\sqrt{T}} \\ \frac{X_1'\varepsilon_{\cdot 1}}{\sqrt{T}} \end{pmatrix} = \begin{pmatrix} \frac{Y_1'X}{T}\left(\frac{X'X}{T}\right)^{-1} \\ I, 0 \end{pmatrix} \frac{X'\varepsilon_{\cdot 1}}{\sqrt{T}}$$

Since

$$\operatorname{plim}\begin{pmatrix} \dfrac{Y_1'X}{T}\left(\dfrac{X'X}{T}\right)^{-1} \\ I,0 \end{pmatrix} = \begin{pmatrix} \Pi_{\cdot 2}'\underline{Q}\,\underline{Q}^{-1} \\ I,0 \end{pmatrix} = \begin{pmatrix} \Pi_{\cdot 2}' \\ I,0 \end{pmatrix}$$

and since $X'\varepsilon_{\cdot 1}/\sqrt{T}$ has asymptotic distribution $N(0,\sigma^2\underline{Q})$, it follows that the asymptotic distribution of

$$\begin{pmatrix} \dfrac{\hat{Y}_1'\varepsilon_{\cdot 1}}{\sqrt{T}} \\ \dfrac{X_1'\varepsilon_{\cdot 1}}{\sqrt{T}} \end{pmatrix}$$

is normal with zero mean and covariance matrix

$$\sigma^2\begin{pmatrix} \Pi_{\cdot 2}' \\ I,0 \end{pmatrix}\underline{Q}\begin{pmatrix} \Pi_{\cdot 2}' \\ I,0 \end{pmatrix}' = \sigma^2\begin{pmatrix} \Pi_{\cdot 2}'\underline{Q}\Pi_{\cdot 2} & \Pi_{\cdot 2}'\underline{Q}_1 \\ \underline{Q}_1'\Pi_{\cdot 2} & \underline{Q}_{11} \end{pmatrix} = \sigma^2\bar{B}$$

It therefore follows that the asymptotic distribution of

$$\sqrt{T}\begin{pmatrix} \hat{\gamma} - \gamma \\ \hat{\delta} - \delta \end{pmatrix}$$

is $N(0,\ {}^2\bar{B}^{-1}\,\bar{B}\,\bar{B}^{-1}{}') = N(0,\sigma^2\bar{B}^{-1})$.

Note that the asymptotic covariance matrix of $\begin{pmatrix} \hat{\gamma} \\ \hat{\delta} \end{pmatrix}_{2SLS}$ is therefore

$$\frac{\sigma^2}{T}\bar{B}^{-1}$$

of which a consistent estimator is

$$s^2\begin{pmatrix} \hat{Y}_1'Y_1 & Y_1'X_1 \\ X_1'Y_1 & X_1'X_1 \end{pmatrix}^{-1}$$

9. PROPOSITION: Let $\hat{\gamma}_i$ and $\hat{\delta}_i$ be the 2SLS estimators of the parameters of the i-th equation. Then the asymptotic distribution of

$$\sqrt{T}\begin{pmatrix} \hat{\gamma}_1 - \gamma_1 \\ \hat{\delta}_1 - \delta_1 \\ \vdots \\ \hat{\gamma}_G - \gamma_G \\ \hat{\delta}_G - \delta_G \end{pmatrix} = \sqrt{T}\ (\hat{\alpha} - \alpha)$$

is $N(0,\overline{B}^*)$, where

$$\overline{B}^* = \begin{pmatrix} \sigma_{11}\overline{B}_{11} & \sigma_{12}\overline{B}_{12} & \cdots & \sigma_{1G}\overline{B}_{1G} \\ \vdots & \vdots & & \vdots \\ \sigma_{G1}\overline{B}_{G1} & \sigma_{G2}\overline{B}_{G2} & \cdots & \sigma_{GG}\overline{B}_{GG} \end{pmatrix}$$

and where

$$\overline{B}_{ij} = \operatorname{plim}\begin{pmatrix} \frac{Y_i'X}{T}\left(\frac{X'X}{T}\right)^{-1}\frac{X'Y_i}{T} & \frac{Y_i'X_i}{T} \\ \frac{X_i'Y_i}{T} & \frac{X_i'X_i}{T} \end{pmatrix}^{-1}\begin{pmatrix} \frac{Y_i'X}{T}\left(\frac{X'X}{T}\right)^{-1}\frac{X'Y_j}{T} & \frac{Y_i'X_j}{T} \\ \frac{X_i'Y_j}{T} & \frac{X_i'X_j}{T} \end{pmatrix}$$

$$\times \begin{pmatrix} \frac{Y_j'X}{T}\left(\frac{X'X}{T}\right)^{-1}\frac{X'Y_j}{T} & \frac{Y_j'X_j}{T} \\ \frac{X_j'Y_j}{T} & \frac{X_j'X_j}{T} \end{pmatrix}^{-1}$$

(It is assumed that all equations are identified, of course.)

Proof: Defining

$$B_{ii} = \frac{1}{T}\begin{pmatrix} Y_i'X(X'X)^{-1}X'Y_i & Y_iX_i \\ X_i'Y_i & X_i'X_i \end{pmatrix}$$

and

$$c_i = \begin{pmatrix} \frac{Y_i'X}{T}\left(\frac{X'X}{T}\right)^{-1} \\ I,0 \end{pmatrix}$$

we have

$$\sqrt{T}\,(\hat{\alpha} - \alpha) = \begin{pmatrix} B_{11}^{-1}c_1 & \dfrac{X'\varepsilon_{\cdot 1}}{\sqrt{T}} \\ \vdots & \vdots \\ B_{GG}^{-1}c_G & \dfrac{X'\varepsilon_{\cdot G}}{\sqrt{T}} \end{pmatrix}$$

(This is essentially the same thing as was done for a single equation in the proof of Theorem 8.) An alternative expression is

$$\sqrt{T}\,(\hat{\alpha} - \alpha) = \begin{pmatrix} B_{11}^{-1}c_1 & & 0 \\ & B_{22}^{-1}c_2 & \\ 0 & & \ddots\, B_{GG}^{-1}c_G \end{pmatrix} \mathrm{vec}\left(\frac{X'\varepsilon}{\sqrt{T}}\right)$$

Now, according to Proposition 4 of Section 4.3, the asymptotic distribution of

$$\mathrm{vec}\,\frac{X'\varepsilon}{\sqrt{T}}$$

is $N(0, \Sigma \otimes \underset{\sim}{Q})$. Hence the asymptotic distribution of $\sqrt{T}\,(\hat{\alpha} - \alpha)$ is normal with mean zero and covariance matrix

$$\mathrm{plim}\begin{pmatrix} B_{11}^{-1}c_1 & & 0 \\ & \ddots & \\ 0 & & B_{GG}^{-1}c_G \end{pmatrix}\begin{pmatrix} \sigma_{11}\underset{\sim}{Q} & \cdots & \sigma_{1G}\underset{\sim}{Q} \\ \vdots & & \vdots \\ \sigma_{G1}\underset{\sim}{Q} & \cdots & \sigma_{GG}\underset{\sim}{Q} \end{pmatrix}\begin{pmatrix} c_1'B_{11}^{-1'} & & 0 \\ & \ddots & \\ 0 & & c_G'B_{GG}^{-1'} \end{pmatrix}$$

$$= \mathrm{plim}\begin{pmatrix} \sigma_{11}B_{11}^{-1}c_1\underset{\sim}{Q}c_1'B_{11}^{-1'} & \cdots & \sigma_{1G}B_{11}^{-1}c_1\underset{\sim}{Q}c_G'B_{GG}^{-1'} \\ \vdots & & \vdots \\ \sigma_{G1}B_{GG}^{-1}c_G\underset{\sim}{Q}c_1'B_{11}^{-1'} & \cdots & \sigma_{GG}B_{GG}^{-1}c_G\underset{\sim}{Q}c_G'B_{GG}^{-1'} \end{pmatrix}$$

To show that this equals $\overline{B}^*$ as defined above, note that the i,j-th block element of this matrix is

$$\text{plim } \sigma_{ij} B_{ii}^{-1} c_i Q c_j' B_{jj}^{-1'}$$

$$= \text{plim } \sigma_{ij} \begin{pmatrix} \frac{\hat{Y}_i' Y_i}{T} & \frac{Y_i' X_i}{T} \\ \frac{X_i' Y_i}{T} & \frac{X_i' X_i}{T} \end{pmatrix}^{-1} \begin{pmatrix} \frac{Y_i' X}{T} \left(\frac{X'X}{T}\right)^{-1} \\ I,0 \end{pmatrix} \left(\frac{X'X}{T}\right)$$

$$\times \begin{pmatrix} \left(\frac{X'X}{T}\right)^{-1} \frac{X'Y_j}{T} & \begin{matrix} I \\ 0 \end{matrix} \end{pmatrix} \begin{pmatrix} \frac{\hat{Y}_j' Y_j}{T} & \frac{Y_j' X_j}{T} \\ \frac{X_j' Y_j}{T} & \frac{X_j' X_j}{T} \end{pmatrix}$$

$$= \sigma_{ij} \overline{B}_{ij}$$

as defined above. But $\overline{B}_{ij}$ is the j-th block element of $\overline{B}^*$.

2SLS and Identification

10. PROPOSITION: The 2SLS estimator is undefined (for any sample size) if the order condition for identification (by the exclusion restrictions) is not met.

Proof: In order for the 2SLS estimator to exist, we need the matrix

$$\begin{pmatrix} Y_1'X(X'X)^{-1}X'Y_1 & Y_1'X_1 \\ X_1'Y_1 & X_1'X_1 \end{pmatrix}$$

to be of rank $g_* + k_* - 1$. But noting that

$$Y_1'X(X'X)^{-1}X'Y_1 = Y_1'X(X'X)^{-1}X' \cdot X(X'X)^{-1}X'Y_1$$

and that $X_1'Y_1 = X_1'X(X'X)^{-1}X'Y_1$ (by Lemma 2), we can write the above matrix as

$$[X(X'X)^{-1}X'Y_1, X_1]'[X(X'X)^{-1}X'Y_1, X_1]$$

Hence its rank is equal to the rank of

$$[X(X'X)^{-1}X'Y_1, X_1] = X[(X'X)^{-1}X'Y_1, \binom{I}{0}]$$

which shows that its rank cannot exceed the rank of X, which is K. Thus for the necessary inverse to exist, we must require

$$g_* + k_* - 1 \le K$$

or

$$g_* - 1 \le k_{**}$$

or

$$G - 1 \le k_{**} + g_{**}$$

which is just the order condition.

11. COMMENT: If the order condition is satisfied but the rank condition is not (that is, if the equation is underidentified even though the order condition is satisfied), the matrix

$$\begin{pmatrix} \hat{Y}_1'Y_1 & Y_1'X_1 \\ X_1'Y_1 & X_1'X_1 \end{pmatrix}$$

will be nonsingular, so that the 2SLS estimator will exist.

It will *not* be a consistent estimator, however, since

$$\operatorname{plim} \frac{1}{T}\begin{pmatrix} \hat{Y}_1'Y_1 & Y_1'X_1 \\ X_1'Y_1 & X_1'X_1 \end{pmatrix}$$

will be singular.

12. PROPOSITION: If the equation is exactly identified, 2SLS and ILS are (numerically) identical.

Proof: ILS is defined by

$$\begin{pmatrix} \hat{\Pi}_{11} & \hat{\Pi}_{12} & \hat{\Pi}_{13} \\ \hat{\Pi}_{21} & \hat{\Pi}_{22} & \hat{\Pi}_{23} \end{pmatrix}\begin{pmatrix} -1 \\ \hat{\gamma} \\ 0 \end{pmatrix} = \begin{pmatrix} -\hat{\delta} \\ 0 \end{pmatrix}$$

or

$$(X'X)^{-1}X'(y,Y_1)\begin{pmatrix} -1 \\ \hat{\gamma} \end{pmatrix} = \begin{pmatrix} -\hat{\delta} \\ 0 \end{pmatrix}$$

or

$$-(X'X)^{-1}X'y + (X'X)^{-1}X'Y_1\hat{\gamma} = \begin{pmatrix} -\hat{\delta} \\ 0 \end{pmatrix}$$

Premultiplying this set of equations by $Y_1'X$ gives

$$-Y_1'X(X'X)^{-1}X'y + Y_1'X(X'X)^{-1}X'Y_1\hat{\gamma} = Y_1'(-X_1\hat{\delta} - X_1^*0)$$

or

$$[\hat{Y}_1'Y_1 \quad Y_1'X_1]\begin{pmatrix} \hat{\gamma} \\ \hat{\delta} \end{pmatrix} = \hat{Y}_1'y$$

On the other hand, premultiplying by $X_1'X$ gives

$$-X_1X(X'X)^{-1}X'y + X_1'X(X'X)^{-1}X'Y_1\hat{\gamma} = X_1'(-X_1\hat{\delta} - X_1^*0)$$

Using the fact that $X_1'X(X'X)^{-1} = [I,0]$, we have

$$-X_1'y + X_1'Y_1\hat{\gamma} = -X_1'X_1\hat{\delta}$$

or

$$[X_1'Y_1 \quad X_1'X_1]\begin{pmatrix} \hat{\gamma} \\ \hat{\delta} \end{pmatrix} = X_1'y$$

Combining this with the above result, we get

$$\begin{pmatrix} \hat{Y}_1'Y_1 & Y_1'X_1 \\ X_1'Y_1 & X_1'X_1 \end{pmatrix}\begin{pmatrix} \hat{\gamma} \\ \hat{\delta} \end{pmatrix} = \begin{pmatrix} \hat{Y}_1'y \\ X_1'y \end{pmatrix}$$

the solution of which is indeed 2SLS.

13. COMMENT: Note that the number of equations characterizing ILS is K, while the number for 2SLS is $g_* + k_* - 1$. In the case of exact identification, $K = g_* + k_* - 1$, and indeed the transformation of the ILS equations into the 2SLS equations is nonsingular, so that the content of the two sets of equations is identical. On the other hand, in the overidentified case the 2SLS equations are less in number than the ILS equations and the two estimators are not identical; indeed, the ILS estimator is gotten by *ignoring* $K - (g_* + k_* - 1)$ of the ILS equations.

Alternative Derivations of 2SLS

14. PROPOSITION: The OLS regression of y on $\hat{Y}_1$ and X_1 yields the 2SLS estimator of γ and δ.

Proof: The OLS estimator is

$$\begin{pmatrix} \hat{\gamma} \\ \hat{\delta} \end{pmatrix} = \begin{pmatrix} \hat{Y}_1'\hat{Y}_1 & \hat{Y}_1'X_1 \\ X_1'\hat{Y}_1 & X_1'X_1 \end{pmatrix}^{-1} \begin{pmatrix} \hat{Y}_1 y \\ X_1 y \end{pmatrix}$$

But since $\hat{Y}_1'X_1 = Y_1'X_1$ (by Lemma 2) and since $\hat{Y}_1'\hat{Y}_1 = \hat{Y}_1'Y_1$, we have

$$\begin{pmatrix} \hat{\gamma} \\ \hat{\delta} \end{pmatrix} = \begin{pmatrix} \hat{Y}_1'Y_1 & Y_1'X_1 \\ X_1'Y_1 & X_1'X_1 \end{pmatrix}^{-1} \begin{pmatrix} \hat{Y}_1'y \\ X_1'y \end{pmatrix} = \begin{pmatrix} \hat{\gamma} \\ \hat{\delta} \end{pmatrix}_{2SLS}$$

In this context, however, it is perhaps worth noting explicitly that the variance estimator associated with the OLS regression of y on $\hat{Y}_1$ and X_1 is not a consistent estimator.

15. PROPOSITION: The estimator

$$\hat{\sigma}^2 = \frac{1}{T}(y - \hat{Y}_1\hat{\gamma} - X_1\hat{\delta})'(y - \hat{Y}_1\hat{\delta} - X_1\hat{\delta})$$

is an inconsistent estimator of σ^2 (the variance of $\varepsilon_{\cdot 1}$).

Proof: Expanding $\hat{\sigma}^2$,

$$\begin{aligned} \hat{\sigma}^2 = \frac{1}{T}[\varepsilon_{\cdot 1} + (Y_1\gamma - \hat{Y}_1\hat{\gamma}) + X_1(\delta - \hat{\delta})]'[\varepsilon_{\cdot 1} + (Y_1\gamma - \hat{Y}_1\hat{\gamma}) \\ + X_1(\delta - \hat{\delta})] \end{aligned}$$

$$\begin{aligned} &= \frac{1}{T}[\varepsilon_{\cdot 1}'\varepsilon_{\cdot 1} + (Y_1\gamma - \hat{Y}_1\hat{\gamma})'(Y_1\gamma - \hat{Y}_1'\hat{\gamma}) + (\delta - \hat{\delta})'X_1'X_1(\delta - \hat{\delta}) \\ &+ 2(\delta - \hat{\delta})X_1'(Y_1\gamma - \hat{Y}_1\hat{\gamma}) + 2(\delta - \hat{\delta})'X_1'\varepsilon_{\cdot 1} \\ &+ 2(Y_1\gamma - \hat{Y}_1\hat{\gamma})'\varepsilon_{\cdot 1}] \end{aligned}$$

It is clear that

$$\operatorname{plim} \frac{1}{T} \varepsilon_{\cdot 1}' \varepsilon_{\cdot 1} = \sigma^2$$

$$\operatorname{plim}(\delta - \hat{\delta})' \frac{X_1' X_1}{T} (\delta - \hat{\delta}) = 0$$

$$\operatorname{plim}(\delta - \hat{\delta})' \frac{X_1'(Y_1\gamma - \hat{Y}_1\hat{\gamma})}{T} = 0$$

$$\operatorname{plim}(\delta - \hat{\delta})' \frac{X_1'\varepsilon_{\cdot 1}}{T} = 0$$

Furthermore,

$$\operatorname{plim} \frac{1}{T} (Y_1\gamma - \hat{Y}_1\hat{\gamma})'(Y_1\gamma - \hat{Y}_1\hat{\gamma})$$

$$= \operatorname{plim}\left[\frac{1}{T} \gamma' Y_1' Y_1 \gamma + \frac{1}{T} \hat{\gamma}' \hat{Y}_1' \hat{Y}_1 \hat{\gamma} - \frac{2}{T} \hat{\gamma}' \hat{Y}_1' Y_1 \gamma\right]$$

$$= \gamma' \operatorname{plim} \frac{Y_1' Y_1}{T} \gamma - \gamma' \operatorname{plim}\left[\frac{Y_1' X}{T} \left(\frac{X'X}{T}\right)^{-1} \frac{X' Y_1}{T}\right]\gamma$$

$$\text{(since } \operatorname{plim} \hat{\gamma} = \gamma \text{ and } \hat{Y}_1'\hat{Y}_1 = \hat{Y}_1' Y_1)$$

$$= \gamma'[\Pi_{\cdot 2}' Q \Pi_{\cdot 2} + \Omega_{**}]\gamma - \gamma' \Pi_{\cdot 2}' Q\, Q^{-1} Q \Pi_{\cdot 2}\gamma$$

$$= \gamma' \Omega_{**} \gamma$$

(Recall that Ω_{**} is the submatrix of Ω corresponding to Y_1, or V_1.) Also,

$$\operatorname{plim} \frac{2}{T} (Y_1\gamma - \hat{Y}_1\hat{\gamma})'\varepsilon_{\cdot 1} = 2\gamma' \operatorname{plim} \frac{Y_1'\varepsilon_{\cdot 1}}{T}$$

$$= -2\gamma'(\Gamma_{**}^{-1})'\Sigma_{\cdot 1}$$

where Γ_{**}^{-1} is the $G \times (g_* - 1)$ submatrix of Γ^{-1} whose columns correspond to Y_1 and $\Sigma_{\cdot 1}$ is the first column of Σ. Therefore,

$$\operatorname{plim} \hat{\sigma}^2 = \sigma^2 + \gamma'\Omega_{**}\gamma - 2\gamma'(\Gamma_{**}^{-1})'\Sigma_{\cdot 1} \neq \sigma^2 \qquad \text{(in general)}$$

16. PROPOSITION: Let the equations

$$y = Y_1\gamma + X_1\delta + \varepsilon_{\cdot 1}$$

be written as

$$y = Z_1\beta + \varepsilon_{\cdot 1} \qquad Z = (Y_1, X_1) \qquad \beta = \begin{pmatrix} \gamma \\ \delta \end{pmatrix}$$

Then

$$\hat\beta = [Z_1'X(X'X)^{-1}X'Z_1]^{-1}Z_1'X(X'X)^{-1}X'y$$

is the 2SLS estimator.

Proof: Note that

$$\hat\beta = \begin{pmatrix} Y_1'X(X'X)^{-1}X'Y_1 & Y_1'X(X'X)^{-1}X'X_1 \\ X_1'X(X'X)^{-1}X'Y_1 & X_1'X(X'X)^{-1}X'X_1 \end{pmatrix}^{-1} \begin{pmatrix} Y_1'X(X'X)^{-1}X'y \\ X_1'X(X'X)^{-1}X'y \end{pmatrix}$$

But since

$$(X'X)^{-1}X'X_1 = \begin{pmatrix} I \\ 0 \end{pmatrix}$$

we have $X(X'X)^{-1}X'X_1 = X_1$. Therefore

$$Y_1'X(X'X)^{-1}X'X_1 = Y_1'X_1$$
$$X_1'X(X'X)^{-1}X'X_1 = X_1'X_1$$
$$X_1'X(X'X)^{-1}X'y = X_1'y$$

With these substitutions,

$$\hat\beta = \begin{pmatrix} Y_1'X(X'X)^{-1}X'Y_1 & Y_1'X_1 \\ X_1'Y_1 & X_1'X_1 \end{pmatrix}^{-1} \begin{pmatrix} Y_1'X(X'X)^{-1}X'y \\ X_1'y \end{pmatrix} = \begin{pmatrix} \hat\gamma \\ \hat\delta \end{pmatrix}_{2SLS}$$

17. COMMENT: Note that if we transform the equation $y = Z_1\beta + \varepsilon_{\cdot 1}$ by X' to get $X'y = X'Z_1\beta + X'\varepsilon_{\cdot 1}$, and then apply GLS using $\sigma^2X'X$ as the covariance matrix of $X'\varepsilon_{\cdot 1}$, the above (2SLS) estimator results.

18. LEMMA: Let A be a positive definite $n \times n$ matrix and B be a positive semidefinite $n \times n$ matrix. Then there exists a nonsingular matrix W such that

$$A = WW' \qquad B = W\Lambda W'$$

where Λ = diagonal $(\lambda_1, \lambda_2, \cdots, \lambda_n)$ and where λ_i are the roots of the determinental equation

$$|B - \lambda A| = 0$$

Proof: Since A is positive definite, there exists a nonsingular matrix T such that $A = TT'$. But then the roots of $|B - \lambda A| = 0$, or of $|B - \lambda TT'| = 0$ are the roots of $|T^{-1}BT'^{-1} - \lambda I||T||T'| = 0$; that is, they are the eigenvalues of $T^{-1}BT'^{-1}$.

Then let D be the matrix of (orthogonalized) eigenvectors of $T^{-1}BT'^{-1}$; by definition, $T^{-1}BT'^{-1}D = D\Lambda$. Therefore $T^{-1}BT'^{-1} = D\Lambda D'$ and $B = TD\Lambda D'T'$.

Finally, let $W = TD$. Then

$$WW' = TDD'T' = TT' = A$$
$$W\Lambda W' = TD\Lambda D'T' = B$$

19. LEMMA: Let X be a $T \times K$ matrix of rank K, and let A be any matrix with K rows and r columns, with $r \le K$ and rank(A) = r. Then

$$(X'X)^{-1} - A(A'X'XA)^{-1}A'$$

is positive semidefinite.

Proof: Since $(X'X)^{-1}$ is positive definite and $A(A'X'XA)^{-1}A'$ is positive semidefinite, by Lemma 18 there exists a nonsingular W such that

$$(X'X)^{-1} = WW'$$
$$A(A'X'XA)^{-1}A' = W\Lambda W'$$

where Λ is the diagonal matrix of roots to the equation

$$|A(A'X'XA)^{-1}A' - \lambda(X'X)^{-1}| = 0$$

Therefore

$$(X'X)^{-1} - A(A'X'XA)^{-1}A' = W(I-\Lambda)W'$$

Now the roots of

$$|A(A'X'XA)^{-1}A' - \lambda(X'X)^{-1}| = 0$$

are the roots of

$$|A(A'X'XA)^{-1}A'X'X - \lambda I| = 0$$

that is, the eigenvalues of $A(A'X'XA)^{-1}A'X'X$. This matrix is of dimension K but only of rank r, so it has K - r zero eigenvalues. Furthermore, the nonzero eigenvalues of $A(A'X'XA)^{-1}A'X'X$ are the same as the (nonzero) eigenvalues of $(A'X'XA)^{-1}A'X'XA = I_r$. Therefore, we have

$$\Lambda = \begin{pmatrix} I_r & 0 \\ 0 & 0 \end{pmatrix} \qquad I_K - \Lambda = \begin{pmatrix} 0 & 0 \\ 0 & I_{K-r} \end{pmatrix}$$

and

$$(X'X)^{-1} - A(A'X'XA)^{-1}A' = W\begin{pmatrix} 0 & 0 \\ 0 & I_{K-r} \end{pmatrix}W'$$

which is indeed positive semidefinite.

20. PROPOSITION: In the model

$$y = Y_1\gamma + X_1\delta + \varepsilon_{\cdot 1} = Z_1\beta + \varepsilon_{\cdot 1}$$

consider the class of instrumental variables estimators

$$\hat{\beta}_{IV} = (P'Z_1)^{-1}P'y$$

where the instruments are of the form $P = XA$ (that is, the instruments are linear combinations of the predetermined variables). Then the most asymptotically efficient such instrumental variables estimator is the 2SLS estimator.

Proof: First note that the 2SLS estimator is in this class; its matrix of instruments is

$$(\hat{Y}_1, X_1) = X\left[(X'X)^{-1}X'Y_1, \binom{I}{0}\right]$$

Also note that, with instruments of the form $P = XA$, it is clear that $\sqrt{T}(\hat{\beta}_{IV} - \beta)$ has asymptotic distribution $N(0, \sigma^2\Psi_{IV})$, where

$$\Psi_{IV} = \text{plim}\left(\frac{P'Z_1}{T}\right)^{-1} \frac{P'P}{T} \left(\frac{Z_1'P}{T}\right)^{-1}$$

For the 2SLS estimator we have the asymptotic distribution of $N(0,\sigma^2\Psi_{2SLS})$, where

$$\Psi_{2SLS} = \text{plim}\left[\frac{Z_1'X}{T}\left(\frac{X'X}{T}\right)^{-1}\frac{X'Z_1}{T}\right]^{-1}$$

This is just the result of Theorem 8 put in the notation of Proposition 16. We will therefore prove the proposition by showing that

$$\left(\frac{P'Z_1}{T}\right)^{-1} \frac{P'P}{T} \left(\frac{Z_1'P}{T}\right)^{-1} - \left[\frac{Z_1'X}{T}\left(\frac{X'X}{T}\right)^{-1}\frac{X'Z_1}{T}\right]^{-1}$$

is positive semidefinite for all T and for any $P = XA$.

Noting that T can be ignored, and using the fact that $A - B$ is positive semidefinite if and only if $B^{-1} - A^{-1}$ is, we need to show that

$$Z_1'X(X'X)^{-1}X'Z_1 - Z_1'P(P'P)^{-1}P'Z_1$$

is positive semidefinite. But substituting $P = XA$, we get

$$Z_1'X[(X'X)^{-1} - A(A'X'XA)^{-1}A']XZ_1$$

and the result follows from Lemma 19.

21. COMMENT: Note that A has K rows and $g_* + k_* - 1$ columns. If the equation is exactly identified, A is square (and nonsingular, by assumption), and the IV estimator is invariant with respect to A. It is only in the overidentified case that the choice of A matters.

Note also that, since the indirect least squares estimator has been shown to be an instrumental variables estimator, with instruments of the form $P = XA$, it follows that two-stage least squares is efficient relative to indirect least squares.

4.7 THE k-CLASS

1. DEFINITION: The k-class estimator of γ and δ in the equation $y = Y_1\gamma + X_1\delta + \varepsilon_{\cdot 1}$ is

$$\begin{pmatrix} \hat{\gamma} \\ \hat{\delta} \end{pmatrix}_k = \begin{pmatrix} Y_1'Y_1 - k\hat{V}_1'\hat{V}_1 & Y_1'X_1 \\ X_1'Y_1 & X_1'X_1 \end{pmatrix}^{-1} \begin{pmatrix} (Y_1 - k\hat{V}_1)'y \\ X_1'y \end{pmatrix}$$

where $\hat{V}_1 = Y_1 - \hat{Y}_1 = Y_1 - X(X'X)^{-1}X'Y_1$.

2. PROPOSITION: The OLS estimator is the k-class estimator corresponding to $k = 0$.

3. PROPOSITION: The 2SLS estimator is the k-class estimator corresponding to $k = 1$.

Proof: Note that $\hat{Y}_1 = Y_1 - \hat{V}_1$. Therefore

$$(Y_1 - (1)\hat{V}_1)'y = \hat{Y}_1'y$$

Also

$$\begin{aligned} Y_1'Y_1 - (1)\hat{V}_1'\hat{V}_1 &= Y_1'Y_1 - (Y_1 - \hat{Y}_1)'(Y_1 - \hat{Y}_1) \\ &= Y_1'Y_1 - Y_1'Y_1 + 2\hat{Y}_1'Y_1 - \hat{Y}_1'\hat{Y}_1 \\ &= \hat{Y}_1'Y_1 \end{aligned}$$

since $\hat{Y}_1'Y_1 = Y_1'\hat{Y}_1 = \hat{Y}_1'\hat{Y}_1$. Therefore

$$\begin{pmatrix} \hat{\gamma} \\ \hat{\delta} \end{pmatrix}_1 = \begin{pmatrix} \hat{Y}_1'Y_1 & Y_1'X_1 \\ X_1'Y_1 & X_1'X_1 \end{pmatrix}^{-1} \begin{pmatrix} \hat{Y}_1'y \\ X_1'y \end{pmatrix} = \begin{pmatrix} \hat{\gamma} \\ \hat{\delta} \end{pmatrix}_{2SLS}$$

4. PROPOSITION: The k-class estimator is an IV estimator, with $Y_1 - k\hat{V}_1$ serving as instrument for Y_1 and with X_1 serving as its own instrument.

Proof: Note that

$$(Y_1 - k\hat{V}_1)'X_1 = Y_1'X_1 \qquad (\text{since } \hat{V}_1'X = 0)$$

Also

$$\begin{aligned}(Y_1 - k\hat{V}_1)'Y_1 &= Y_1'Y_1 - k\hat{V}_1'Y_1 \\ &= Y_1'Y_1 - k\hat{V}_1'\hat{V}_1 - k\hat{V}_1'\hat{Y}_1 \qquad (\text{since } Y_1 = \hat{Y}_1 + \hat{V}_1) \\ &= Y_1'Y_1 - k\hat{V}_1'\hat{V}_1 \qquad (\text{since } \hat{V}_1'\hat{Y}_1 = 0)\end{aligned}$$

Then the IV estimator is

$$\begin{pmatrix}\hat{\gamma}\\ \hat{\delta}\end{pmatrix} = \begin{pmatrix}(Y_1 - k\hat{V}_1)'Y_1 & (Y_1 - k\hat{V}_1)'X_1\\ X_1'Y_1 & X_1'X_1\end{pmatrix}^{-1}\begin{pmatrix}(Y_1 - k\hat{V}_1)'y\\ X_1'y\end{pmatrix}$$

$$= \begin{pmatrix}Y_1'Y_1 - k\hat{V}_1'\hat{V}_1 & Y_1'X_1\\ X_1'Y_1 & X_1'X_1\end{pmatrix}^{-1}\begin{pmatrix}(Y_1 - k\hat{V}_1)'y\\ X_1'y\end{pmatrix}$$

$$= \begin{pmatrix}\hat{\gamma}\\ \hat{\delta}\end{pmatrix}_k$$

5. THEOREM: The k-class estimator is consistent if plim $k = 1$; it has the same asymptotic covariance matrix as 2SLS if plim $\sqrt{T}(k - 1) = 0$ (if the equation is identified, of course).

Proof: First note that

$$\begin{pmatrix}\hat{\gamma}\\ \hat{\delta}\end{pmatrix}_k = \begin{pmatrix}\gamma\\ \delta\end{pmatrix} + \begin{pmatrix}Y_1'Y_1 - k\hat{V}_1'\hat{V}_1 & Y_1'X_1\\ X_1'Y_1 & X_1'X_1\end{pmatrix}^{-1}\begin{pmatrix}(Y_1 - k\hat{V}_1)'\varepsilon_{\cdot 1}\\ X_1'\varepsilon_{\cdot 1}\end{pmatrix}$$

Hence

$$\text{plim}\begin{pmatrix}\hat{\gamma}\\ \hat{\delta}\end{pmatrix}_k = \begin{pmatrix}\gamma\\ \delta\end{pmatrix} + \text{plim}\begin{pmatrix}\dfrac{Y_1'Y_1 - k\hat{V}_1'\hat{V}_1}{T} & \dfrac{Y_1'X_1}{T}\\ \dfrac{X_1'Y_1}{T} & \dfrac{X_1'X_1}{T}\end{pmatrix}^{-1}\text{plim}\begin{pmatrix}\dfrac{(Y_1 - k\hat{V}_1)'\varepsilon_{\cdot 1}}{T}\\ \dfrac{X_1'\varepsilon_{\cdot 1}}{T}\end{pmatrix}$$

Now, if plim k = 1,

$$\text{plim}\,\frac{Y_1'Y_1 - k\hat{V}_1'\hat{V}_1}{T} = \text{plim}\,\frac{Y_1'Y_1 - \hat{V}_1'\hat{V}_1}{T} = \text{plim}\,\frac{\hat{Y}_1'Y_1}{T}$$

Therefore

$$\text{plim}\begin{pmatrix} \dfrac{Y_1'Y_1 - k\hat{V}_1'\hat{V}_1}{T} & \dfrac{Y_1'X_1}{T} \\ \dfrac{X_1'Y_1}{T} & \dfrac{X_1'X_1}{T} \end{pmatrix}^{-1} = \text{plim}\begin{pmatrix} \dfrac{\hat{Y}_1'Y_1}{T} & \dfrac{Y_1'X_1}{T} \\ \dfrac{X_1'Y_1}{T} & \dfrac{X_1'X_1}{T} \end{pmatrix}^{-1}$$

Furthermore, this matrix is finite, and nonsingular if the equation is identified. Hence to prove consistency all that remains is to show that

$$\text{plim}\begin{pmatrix} \dfrac{(Y_1 - k\hat{V}_1)'\varepsilon_{\cdot 1}}{T} \\ \dfrac{X_1'\varepsilon_{\cdot 1}}{T} \end{pmatrix} = 0$$

It is clear that $\text{plim}(X_1'\varepsilon_{\cdot 1}/T) = 0$. Also

$$\text{plim}\left(\frac{(Y_1 - k\hat{V}_1)'\varepsilon_{\cdot 1}}{T}\right) = \text{plim}(1 - k)\left(\frac{Y_1'\varepsilon_{\cdot 1}}{T}\right) + \text{plim}\,k\left(\frac{\hat{Y}_1'\varepsilon_{\cdot 1}}{T}\right)$$

This equals zero since $\text{plim}(1 - k) = 0$ and since $\text{plim}(\hat{Y}_1'\varepsilon_{\cdot 1}/T) = 0$. Thus the first part of the theorem is proved.

To show the second part of the theorem, note that

$$\sqrt{T}\left[\begin{pmatrix} \hat{\gamma} \\ \hat{\delta} \end{pmatrix}_k - \begin{pmatrix} \gamma \\ \delta \end{pmatrix}\right] = \begin{pmatrix} \dfrac{Y_1'Y_1 - k\hat{V}_1'\hat{V}_1}{T} & \dfrac{Y_1'X_1}{T} \\ \dfrac{X_1'Y_1}{T} & \dfrac{X_1'X_1}{T} \end{pmatrix} \begin{pmatrix} \dfrac{(Y_1 - k\hat{V}_1)'\varepsilon_{\cdot 1}}{\sqrt{T}} \\ \dfrac{X_1'\varepsilon_{\cdot 1}}{\sqrt{T}} \end{pmatrix}$$

Now, if $\text{plim}\,\sqrt{T}(k - 1) = 0$, certainly plim k = 1. But then we have just shown above that the inverse matrix on the right-hand side is identical (in probability limit) to the comparable matrix for the 2SLS estimator. Also, the term

$$\frac{X_1'\varepsilon_{\cdot 1}}{\sqrt{T}}$$

arises in both the k-class and 2SLS estimators. So all that remains is to show that

$$\frac{(Y_1 - k\hat{V}_1)'\varepsilon_{\cdot 1}}{\sqrt{T}} \quad \text{and} \quad \frac{\hat{Y}_1'\varepsilon_{\cdot 1}}{\sqrt{T}}$$

have the same asymptotic distribution.

To do this, it is sufficient to show that

$$\text{plim } \frac{1}{\sqrt{T}} (Y_1 - k\hat{V}_1 - \hat{Y}_1)'\varepsilon_{\cdot 1} = 0$$

But

$$\text{plim } \frac{1}{\sqrt{T}} (Y_1 - k\hat{V}_1 - \hat{Y}_1)'\varepsilon_{\cdot 1}$$

$$= \text{plim } \sqrt{T}\,(1 - k) \frac{Y_1'\varepsilon_{\cdot 1}}{T} - \text{plim } \sqrt{T}\,(1 - k) \frac{\hat{Y}_1'\varepsilon_{\cdot 1}}{T}$$

which equals zero if $\text{plim } \sqrt{T}(1 - k) = 0$. This proves the second part of the theorem.

4.8 LIMITED INFORMATION MAXIMUM LIKELIHOOD

In this section we will again consider the first equation in a system of simultaneous equations, written as

$$y = Y_1\gamma + X_1\delta + \varepsilon_{\cdot 1}$$

However, rather than impose the normalization which is implicit in the above equation, let us define

$$Y_1^o = (y, Y_1)$$

$$\gamma^o = \begin{pmatrix} -1 \\ \gamma \end{pmatrix}$$

and write the equation as

$$Y_1^o\gamma^o + X_1\delta + \varepsilon_{\cdot 1} = 0$$

Least Variance Ratio Estimation

1. DEFINITION: The *least variance ratio* (LVR) estimator $\hat{\gamma}^o$ is the value which minimizes the variance ratio

$$\ell = \frac{\gamma^{o\prime}Y_1^{o\prime}M_1Y_1^o\gamma^o}{\gamma^{o\prime}Y_1^{o\prime}MY_1^o\gamma^o}$$

where $M_1 = I - X_1(X_1'X_1)^{-1}X_1'$ and $M = I - X(X'X)^{-1}X'$. The LVR estimator of δ is $\hat{\delta} = -(X_1'X_1)^{-1}X_1'Y_1^o\hat{\gamma}^o$.

2. COMMENT: $\gamma^{o\prime}Y_1^{o\prime}M_1Y_1^o\gamma^o$ is the sum of squared errors when the artificial variable $Y_1^o\gamma^o$ is regressed on X_1; $\gamma^{o\prime}Y_1^{o\prime}MY_1^o\gamma^o$ is the sum of squared errors when $Y_1^o\gamma^o$ is regressed on $X = (X_1, X_1^*)$. It is therefore clear that $\ell \geq 1$.

3. THEOREM: The LVR estimator $\hat{\gamma}^o$ is the solution to the equation

$$(Y_1^{o\prime}M_1Y_1^o - \hat{\ell}Y_1^{o\prime}MY_1^o)\hat{\gamma}^o = 0$$

where $\hat{\ell}$ is the smallest root of the determinental equation

$$|Y_1^{o\prime}M_1Y_1^o - \hat{\ell}Y_1^{o\prime}MY_1^o| = 0$$

(and is also the minimized value of the variance ratio).

Proof: We wish to minimize the variance ratio ℓ defined in Definition 1.

$$0 = \frac{d\ell}{d\hat{\gamma}^o} = \frac{2Y_1^{o\prime}M_1Y_1^o\hat{\gamma}^o}{\hat{\gamma}^{o\prime}Y_1^{o\prime}MY_1^o\hat{\gamma}^o} - \frac{\hat{\gamma}^{o\prime}Y_1^{o\prime}M_1Y_1^o\hat{\gamma}^o}{(\hat{\gamma}^{o\prime}Y_1^{o\prime}MY_1^o\hat{\gamma}^o)^2}\,2Y_1^{o\prime}MY_1^o\hat{\gamma}^o$$

[see Note 1]. Multiplying through by $\frac{1}{2}\,\hat{\gamma}^{o\prime}Y_1^{o\prime}MY_1^o\hat{\gamma}^o$ gives

$$(Y_1^{o\prime}M_1Y_1^o - \hat{\ell}Y_1^{o\prime}MY_1^o)\hat{\gamma}^o = 0$$

where

$$\hat{\ell} = \frac{\hat{\gamma}^{o\prime} Y_1^{o\prime} M_1 Y_1^o \hat{\gamma}^o}{\hat{\gamma}^{o\prime} Y_1^{o\prime} M Y_1^o \hat{\gamma}^o}$$

Now, for a solution $\hat{\gamma}^o$ to exist, the matrix $Y_1^{o\prime} M_1 Y_1^o - \hat{\ell} Y_1^{o\prime} M Y_1^o$ must be singular; hence $\hat{\ell}$ is a solution to the equation

$$|Y_1^{o\prime} M_1 Y_1^o - \hat{\ell} Y_1^{o\prime} M Y_1^o| = 0$$

Finally, since we wish to *minimize* the variance ratio, we clearly take the smallest such root.

4. COMMENT: It is clear that some normalization rule must be imposed if the solution to

$$(Y_1^{o\prime} M_1 Y_1^o - \hat{\ell} Y_1^{o\prime} M Y_1^o)\hat{\gamma}^o = 0$$

is to be unique. *Which* normalization rule is imposed is of no consequence.

5. PROPOSITION: The LVR estimator is the k-class estimator corresponding to $k = \hat{\ell}$.

Proof: The k-class estimator with $k = \hat{\ell}$ is defined by

$$\begin{pmatrix} Y_1'Y_1 - \hat{\ell}\hat{V}_1'\hat{V}_1 & Y_1'X_1 \\ X_1'Y_1 & X_1'X_1 \end{pmatrix} \begin{pmatrix} \hat{\gamma} \\ \hat{\delta} \end{pmatrix} = \begin{pmatrix} Y_1'y - \hat{\ell}\hat{V}_1'y \\ X_1'y \end{pmatrix}$$

The first partitioned row of this equation is

$$(Y_1'Y_1 - \hat{\ell}\hat{V}_1'\hat{V}_1)\hat{\gamma} + Y_1'X_1\hat{\delta} = Y_1'y - \hat{\ell}\hat{V}_1'y$$

Letting $\hat{\gamma}^o$ be defined as previously, this can be written

$$[Y_1'y - \hat{\ell}\hat{V}_1'y \quad Y_1'Y_1 - \hat{\ell}\hat{V}_1'V_1]\hat{\gamma}^o + Y_1'X_1\hat{\delta} = 0$$

Now note that $\hat{V}_1'y = Y_1'My = Y_1'MMy = \hat{V}_1'\hat{v}_1$, where $\hat{v}_1 = My$. Hence we have

$$[Y_1'y - \hat{\ell}\hat{V}_1'\hat{v}_1 \quad Y_1'Y_1 - \hat{\ell}\hat{V}_1'\hat{V}_1]\hat{\gamma}^o + Y_1'X_1\hat{\delta} = 0$$

Then let $\hat{V}_1^o = (\hat{v}_1, \hat{V}_1)$ and (as before) $Y_1^o = (y, Y_1)$. With these substitutions we get

$$(Y_1'Y_1^o - \hat{\ell}\hat{V}_1'\hat{V}_1^o)\hat{\gamma}^o + Y_1'X_1\hat{\delta} = 0$$

On the other hand, the second partitioned row of the equation defining the k-class is

$$X_1'Y_1\hat{\gamma} + X_1'X_1\hat{\delta} = X_1'y$$

or

$$\hat{\delta} = -(X_1'X_1)^{-1}X_1'Y_1^o\hat{\gamma}^o$$

This is just the formula for the LVR estimator of δ, given that we can indeed show $\hat{\gamma}^o$ to be the LVR estimator of γ^o. To do so, make this substitution for $\hat{\delta}$ in the equation $(Y_1'Y_1^o - \hat{\ell}\hat{V}_1'\hat{V}_1^o)\hat{\gamma}^o + Y_1'X_1\hat{\delta} = 0$ to get

$$[Y_1'Y_1^o - \hat{\ell}\hat{V}_1'\hat{V}_1^o - Y_1'X_1(X_1'X_1)^{-1}X_1'Y_1^o]\hat{\gamma}^o = 0$$

But $\hat{V}_1'\hat{V}_1^o$ is all but the first row of $Y_1^{o\prime}MY_1^o$, and

$$Y_1'Y_1^o - Y_1'X_1(X_1'X_1)^{-1}X_1'Y_1^o$$

is all but the first row of $Y_1^{o\prime}M_1Y_1^o$. Hence we get

$$(Y_1^{o\prime}M_1Y_1^o - \hat{\ell}Y_1^{o\prime}MY_1^o)_{\text{minus first row}}\,\hat{\gamma}^o = 0$$

Now, the LVR estimator is defined by

$$(Y_1^{o\prime}M_1Y_1^o - \hat{\ell}Y_1^{o\prime}MY_1^o)\hat{\gamma}^o = 0$$

However, the matrix $Y_1^{o\prime}M_1Y_1^o - \hat{\ell}Y_1^{o\prime}MY_1^o$ is singular so that its first row (or any row) is just a linear combination of the other rows and could just as well be dropped since it puts no independent restriction on $\hat{\gamma}^o$. But dropping the first row gives us the expression

$$(Y_1^{o\prime}M_1Y_1^o - \hat{\ell}Y_1^{o\prime}MY_1^o)_{\text{minus first row}}\,\hat{\gamma}^o = 0$$

which was just shown to be the k-class estimator with $k = \hat{\ell}$.

6. THEOREM: The LVR estimator is consistent.

Proof: Since the LVR estimator is the k-class estimator with $k = \hat{\ell}$, to show consistency we must show that plim $\hat{\ell} = 1$. But $\hat{\ell}$ is defined as the smallest root of

$$|Y_1^{o}{}'M_1Y_1^{o} - \hat{\ell}Y_1^{o}{}'MY_1^{o}| = 0$$

or, equivalently, of

$$|[Y_1^{o}{}'(M_1 - M)Y_1^{o}](Y_1^{o}{}'MY_1^{o})^{-1} - (\hat{\ell} - 1)I| = 0$$

Therefore $\hat{\ell} - 1$ is the smallest eigenvalue of

$$[Y_1^{o}{}'(M_1 - M)Y_1^{o}](Y_1^{o}{}'MY_1^{o})^{-1}$$

Since $\hat{\ell} \geq 1$, all the eigenvalues of this matrix must be nonnegative. Therefore we simply need to show that asymptotically one root equals zero; that is, that the probability limit of this matrix is singular.

We therefore consider

$$\text{plim}[\tfrac{1}{T} Y_1^{o}{}'(M_1 - M)Y_1^{o}](\tfrac{1}{T} Y_1^{o}{}'MY_1^{o})^{-1}$$

To show that this is singular, it is sufficient to show that

$$\text{plim}[\tfrac{1}{T} Y_1^{o}{}'(M_1 - M)Y_1^{o}]$$

is singular. But note that

$$\text{plim} \tfrac{1}{T} Y_1^{o}{}'MY_1^{o} = \Omega_0$$

the submatrix of the reduced-form covariance matrix corresponding to Y_1^{o}. Also

$$\begin{aligned}\text{plim} \tfrac{1}{T} Y_1^{o}{}'M_1Y_1^{o} &= \text{plim} \tfrac{1}{T} V_1^{o}{}'M_1V_1^{o} + 2 \text{ plim} \tfrac{1}{T} V_1^{o}{}'M_1X\Pi^{o} \\ &\quad + \text{plim} \tfrac{1}{T} \Pi_{2\cdot}^{o}{}'X_1^{*}{}'M_1X_1^{*}\Pi_{2\cdot}^{o} \\ &= \Omega_0 + \Pi_{2\cdot}^{o}{}'(\text{plim} \tfrac{1}{T} X_1^{*}{}'M_1X_1^{*})\Pi_{2\cdot}^{o}\end{aligned}$$

Therefore

$$\text{plim}[\frac{1}{T} Y_1^{o\prime}(M_1 - M)Y_1^o] = \Pi_{2.}^{o\prime}(\text{plim} \frac{1}{T} X_1^{*\prime} M_1 X_1^*)\Pi_{2.}^o$$

This is a matrix of dimension $g_* \times g_*$, whose rank connot exceed $g_* - 1$, the rank of $\Pi_{2.}^o$. It is therefore singular, thus proving the theorem.

7. THEOREM: The LVR estimator has the same asymptotic distribution as the 2SLS estimator.

Proof: We can prove the theorem by proving that plim $\sqrt{T}(\hat{\ell} - 1) = 0$. But as a matter of definition, we have

$$\sqrt{T}(\hat{\ell} - 1) = \sqrt{T}\left(\frac{\hat{\gamma}^{o\prime} Y_1^{o\prime} M_1 Y_1^o \hat{\gamma}^o}{\hat{\gamma}^{o\prime} Y_1^{o\prime} M Y_1^o \hat{\gamma}^o} - 1 \right)$$

$$= \sqrt{T} \frac{\hat{\gamma}^{o\prime} Y_1^{o\prime}(M_1 - M) Y_1^o \hat{\gamma}^o}{\hat{\gamma}^{o\prime} Y_1^{o\prime} M Y_1^o \hat{\gamma}^o}$$

Now we can note that

$$Y_1^o \hat{\gamma}_0 = -X_1\delta - \varepsilon_{.1} + Y_1^o(\hat{\gamma}^o - \gamma^o)$$

It is therefore clear that

$$\text{plim} \frac{1}{T} \hat{\gamma}^{o\prime} Y_1^{o\prime} M Y_1^o \hat{\gamma}^o = \text{plim} \frac{1}{T} \varepsilon_{.1}^{\prime} M \varepsilon_{.1}$$

$$+ \text{plim}(\hat{\gamma}^o - \gamma^o)' \frac{Y_1^{o\prime} M Y_1^o}{T} (\hat{\gamma}^o - \gamma^o)$$

$$- 2 \text{ plim} \frac{\varepsilon_{.1}^{\prime} M Y_1^o}{T} (\hat{\gamma}^o - \gamma^o)$$

$$= \sigma^2 \qquad \text{(since plim } \hat{\gamma}^o = \gamma^o)$$

As a result we need to consider only

$$\frac{1}{\sqrt{T}} \hat{\gamma}^{o\prime} Y_1^{o\prime}(M_1 - M) Y_1^o \hat{\gamma}^o$$

$$= \frac{1}{\sqrt{T}} [-\varepsilon_{.1} + Y_1^o(\hat{\gamma}^o - \gamma^o)]'(M_1 - M)[-\varepsilon_{.1} + Y_1^o(\hat{\gamma}^o - \gamma^o)]$$

$$= \frac{1}{\sqrt{T}} \varepsilon_{\cdot 1}'(M_1 - M)\varepsilon_{\cdot 1} + \sqrt{T}(\hat{\gamma}^o - \gamma^o)' \frac{Y_1^{o\prime}(M_1 - M)Y_1^o}{T} (\hat{\gamma}^o - \gamma^o)$$

$$- 2\sqrt{T}(\hat{\gamma}^o - \gamma^o)'Y_1^{o\prime} \frac{(M_1 - M)\varepsilon_{\cdot 1}}{T}$$

Now we know that $M_1 - M$ is idempotent, of rank k_{**} (see the proof of Proposition 16 of Section 1.5). Therefore $\varepsilon_{\cdot 1}'(M_1 - M)\varepsilon_{\cdot 1}$ has a chi-squared distribution with k_{**} degrees of freedom. When divided by $\sqrt{T}$, its probability limit is zero. Therefore the first term above has a probability limit of zero. Also, the second term above has a probability limit of zero since $\text{plim}(\hat{\gamma}^o - \gamma^o) = 0$, and the third term above has a probability limit of zero since

$$\text{plim} \frac{(M_1 - M)\varepsilon_{\cdot 1}}{T} = 0$$

This establishes that

$$\text{plim} \frac{1}{\sqrt{T}} \hat{\gamma}^{o\prime}Y_1^{o\prime}(M_1 - M)Y_1^o\hat{\gamma}^o = 0$$

and hence that

$$\text{plim} \sqrt{T}(\hat{\ell} - 1) = 0$$

We now derive the properties of the LVR estimator when the equation is exactly identified.

8. LEMMA: Let $X = (X_1, X_1^*)$, $M_1 = I - X_1(X_1'X_1)^{-1}X_1'$, and $M = I - X(X'X)^{-1}X'$. Then

$$M_1 - M = M_1X_1^*(X_1^{*\prime}M_1X_1^*)^{-1}X_1^{*\prime}M_1$$

Proof: According to Lemma 14 of Section 1.5,

$$(X'X)^{-1} =$$

$$\begin{pmatrix} (X_1'X_1)^{-1}[I + X_1'X_1^*(X_1^{*\prime}M_1X_1^*)^{-1}X_1^{*\prime}X_1(X_1'X_1)^{-1}] & -(X_1'X_1)^{-1}X_1'X_1^*(X_1^{*\prime}M_1X_1^*)^{-1} \\ -(X_1^{*\prime}M_1X_1^*)^{-1}X_1^{*\prime}X_1(X_1'X_1)^{-1} & (X_1^{*\prime}M_1X_1^*)^{-1} \end{pmatrix}$$

Hence

$$X(X'X)^{-1}X' = X_1(X_1'X_1)^{-1}X_1'$$
$$+ X_1(X_1'X_1)^{-1}X_1'X_1^*(X_1^{*\prime}M_1X_1^*)^{-1}X_1^{*\prime}X_1(X_1'X_1)^{-1}X_1'$$
$$- X_1^*(X_1^{*\prime}M_1X_1^*)^{-1}X_1^{*\prime}X_1(X_1'X_1)^{-1}X_1'$$
$$- X_1(X_1'X_1)^{-1}X_1'X_1^*(X_1^{*\prime}M_1X_1^*)^{-1}X_1^* + X_1^*(X_1^{*\prime}M_1X_1^*)^{-1}X_1^{*\prime}$$
$$= X_1(X_1'X_1)^{-1}X_1'$$
$$+ [I - X_1(X_1'X_1)^{-1}X_1']X_1^*(X_1^{*\prime}M_1X_1^*)^{-1}X_1^{*\prime}[I - X_1(X_1'X_1)^{-1}X_1']$$

and

$$M_1 - M = X(X'X)^{-1}X' - X_1(X_1'X_1)^{-1}X_1' = M_1X_1^*(X_1^{*\prime}M_1X_1^*)^{-1}X_1^{*\prime}M_1$$

9. PROPOSITION: If the equation is exactly identified, $\hat{\ell} = 1$.

Proof: $\hat{\ell}$ is defined as the smallest root of

$$|Y_1^{o\prime}M_1Y_1^o - \hat{\ell}Y_1^{o\prime}MY_1^o| = 0$$

or of

$$|Y_1^{o\prime}M_1Y_1^o - Y_1^{o\prime}MY_1^o - (\hat{\ell} - 1)Y_1^{o\prime}MY_1^o| = 0$$

or of

$$|(Y_1^{o\prime}M_1Y_1^o - Y_1^{o\prime}MY_1^o)(Y_1^{o\prime}MY_1^o)^{-1} - (\hat{\ell} - 1)I| = 0$$

since $Y_1^{o\prime}MY_1^o$ is nonsingular with probability one. Hence $(\hat{\ell} - 1)$ is the smallest eigenvalue of

$$(Y_1^{o\prime}M_1Y_1^o - Y_1^{o\prime}MY_1^o)(Y_1^{o\prime}MY_1^o)^{-1}$$

We know that these roots must be nonnegative since $\hat{\ell} \geq 1$, so that all we need to show is that (at least) one of the roots is zero, that is, that the matrix is singular. To show this, it is sufficient to show that

$$Y_1^{o\prime}M_1Y_1^o - Y_1^{o\prime}MY_1^o$$

is singular. But note that (using Lemma 8)

$$Y_1^{o\prime}M_1Y_1^o - Y_1^{o\prime}MY_1^o = Y_1^{o\prime}M_1X_1^*(X_1^{*\prime}M_1X_1^*)^{-1}X_1^{*\prime}M_1Y_1^o$$

so that its maximum rank is the rank of X_1^*, which is k_{**}. Now if the equation is exactly identified, $k_{**} = g_* - 1$, so that the maximum rank of $Y_1^{o\prime}(M_1 - M)Y_1^o$ is $g_* - 1$. Since its dimension is $g_* \times g_*$, $Y_1^{o\prime}(M_1 - M)Y_1^o$ is indeed singular, and the value of $\hat{\ell} - 1$ is zero, so that $\hat{\ell} = 1$.

10. COROLLARY: If the equation is exactly identified, the LVR estimator is *identical* to the 2SLS (and ILS) estimator.

Proof: The k-class estimator with $k = \hat{\ell} = 1$ is the 2SLS estimator.

Least Generalized Residual Variance Estimation

Again, consider the equation

$$Y_1^o\gamma^o + X_1\delta + \varepsilon_{\cdot 1} = 0$$

The reduced-form equation associated with Y_1^o is

$$(y, Y_1) = X\begin{pmatrix} \Pi_{11} & \Pi_{12} \\ \Pi_{21} & \Pi_{22} \end{pmatrix} + (v_1, V_1)$$

or

$$Y_1^o = X\Pi^o + V_1^o$$

Letting $\Pi_{1\cdot}^o = (\Pi_{11}, \Pi_{12})$ and $\Pi_{2\cdot}^o = [\Pi_{21}, \Pi_{22}]$, we can write

$$Y_1^o = X_1\Pi_{1\cdot}^o + X_1^*\Pi_{2\cdot}^o + V_1^o$$

Now, the reduced form tells us that $\Pi_{2\cdot}^o\gamma^o = 0$. If we estimate $\Pi_{2\cdot}^o$ by OLS, then indirect least squares amounts to solving $\hat{\Pi}_{2\cdot}^o\hat{\gamma}^o = 0$ for $\hat{\gamma}^o$. Of course, if the equation is overidentified, $\hat{\Pi}_{2\cdot}^o$ will be of rank g_* and no solution will exist. On the other hand, if $\Pi_{2\cdot}^o$ were estimated subject to the restriction that its rank is only $g_* - 1$, then the equation $\hat{\Pi}^o_{2.}\hat{\gamma}^o = 0$ would have a solution (unique up to normalization).

Finally, recall that the OLS estimator $\hat{\Pi}^o = (X'X)^{-1}X'Y^o_1$ minimizes the generalized residual variance

$$|(Y^o_1 - X\Pi^o)'(Y^o_1 - X\Pi^o)|$$

What we will now do is to estimate Π^o by minimizing the generalized residual variance subject to the restriction that $\Pi^o_{2.}\gamma^o = 0$ (that is, that $\Pi^o_{2.}$ has rank $g_* - 1$). Given such an estimate $\hat{\Pi}^o_{2.}$, γ^o can be estimated by solving the equation $\hat{\Pi}^o_{2.}\hat{\gamma}^o = 0$ for $\hat{\gamma}^o$.

11. DEFINITION: The least generalized residual variance (LGRV) estimator of γ^o is the estimator gotten by minimizing

$$|(Y^o_1 - X\Pi^o)'(Y^o_1 - X\Pi^o)|$$

subject to the constraint that $\Pi^o_{2.}\gamma^o = 0$. (The notation is that of the proceeding discussion.)

12. THEOREM: The LGRV estimator is *identical* to the LVR estimator.

Proof: Letting $W = (Y^o_1 - X\Pi^o)'(Y^o_1 - X\Pi^o)$, we may as well minimize $\frac{1}{2}\log|W|$. Hence we form the Lagrangian

$$L = \tfrac{1}{2}\log|W| - \lambda'\Pi^o_{2.}\gamma^o$$

Now, the proof of Theorem 13 of Section 2.6 establishes the fact that

$$\frac{\partial \log|W|}{\partial \Pi^o} = -2(X'Y^o_1 - X'X\Pi^o)W^{-1}$$

Also it is clear that

$$\frac{\partial \lambda'\Pi^o_{2.}\gamma^o}{\partial \Pi^o_{.1}} = 0$$

$$\frac{\partial \lambda'\Pi^o_{2.}\gamma^o}{\partial \Pi^o_{2.}} = \lambda\gamma^{o\prime}$$

Hence

$$\frac{\partial\lambda'\Pi^o_{2.}\gamma^o}{\partial\Pi^o} = \begin{pmatrix} 0 \\ \lambda\gamma^{o\prime} \end{pmatrix}$$

We now proceed with the minimization by taking partials with respect to Π^o, λ, and γ^o, and setting them equal to zero.

$$0 = \frac{\partial L}{\partial\Pi^o} = -(X'Y^o_1 - X'X\hat{\Pi}^o)\hat{W}^{-1} - \begin{pmatrix} 0 \\ \lambda\hat{\gamma}^{o\prime} \end{pmatrix} \tag{1}$$

$$0 = \frac{\partial L}{\partial\lambda} = -\hat{\Pi}^o_{2.}\hat{\gamma}^o \tag{2}$$

$$0 = \frac{\partial L}{\partial\gamma^o} = -\hat{\Pi}^o_{2.}{}'\lambda \tag{3}$$

Postmultiplying (1) by $\hat{W}$ and putting it in partitioned form gives

$$X_1'Y^o_1 - X_1'X_1\hat{\Pi}^o_{1.} - X_1'X^*_1\hat{\Pi}^o_{2.} = 0$$

$$X^{*\prime}_1Y^o_1 - X^{*\prime}_1X_1\hat{\Pi}^o_{1.} - X^{*\prime}_1X^*_1\hat{\Pi}^o_{2.} + \lambda\hat{\gamma}^{o\prime}\hat{W} = 0$$

Solving the first of these two equations for $\hat{\Pi}^o_{1.}$ gives

$$\hat{\Pi}^o_{1.} = (X_1'X_1)^{-1}X_1'(Y^o_1 - X^*_1\hat{\Pi}^o_{2.}) \tag{4}$$

Substituting in the second of the two equations, we get

$$X^{*\prime}_1Y^o_1 - X^{*\prime}_1X_1(X_1'X_1)^{-1}X_1'(Y^o_1 - X^*_1\hat{\Pi}^o_{2.}) - X^{*\prime}_1X^*_1\hat{\Pi}^o_{2.} + \lambda\hat{\gamma}^{o\prime}\hat{W} = 0$$

Solving this for $\hat{\Pi}^o_{2.}$,

$$\hat{\Pi}^o_{2.} = (X^{*\prime}_1M_1X^*_1)^{-1}X^{*\prime}_1M_1Y^o_1 + (X^{*\prime}_1M_1X^*_1)^{-1}\lambda\hat{\gamma}^{o\prime}\hat{W} \tag{5}$$

where as before $M_1 = I - X_1(X_1'X_1)^{-1}X_1'$.

Since $\hat{\Pi}^o_{2.}\hat{\gamma}^o = 0$ [from (2)],

$$(X^{*\prime}_1M_1X^*_1)^{-1}X^{*\prime}_1M_1Y^o_1\hat{\gamma}^o + (X^{*\prime}_1M_1X^*_1)^{-1}\lambda\hat{\gamma}^{o\prime}\hat{W}\hat{\gamma}^o = 0$$

Solving this for λ gives

$$\lambda = -(\hat{\gamma}^{o\prime}\hat{W}\hat{\gamma}^{o})^{-1}X_1^{*\prime}M_1Y_1^{o}\hat{\gamma}^{o} \tag{6}$$

(Note that $\hat{\gamma}^{o\prime}\hat{W}\hat{\gamma}^{o}$ is a scalar.) Substituting this back in (5) yields

$$\hat{\Pi}_{2\cdot}^{o} = (X_1^{*\prime}M_1X_1^{*})^{-1}X_1^{*\prime}M_1Y_1^{o}[I - (\hat{\gamma}^{o\prime}\hat{W}\hat{\gamma}^{o})^{-1}\hat{\gamma}^{o}\hat{\gamma}^{o\prime}\hat{W}] \tag{7}$$

Now note that

$$Y_1^{o} - X\hat{\Pi}^{o} = Y_1^{o} - X_1\hat{\Pi}_{1\cdot}^{o} - X_1^{*}\hat{\Pi}_{2\cdot}^{o} = Y_1^{o} - X_1(X_1'X_1)^{-1}X_1'(Y_1^{o} - X_1^{*}\hat{\Pi}_{2}^{o})$$
$$- X_1^{*}\hat{\Pi}_{2\cdot}^{o} \qquad \text{[using (4)]}$$

so that

$$Y_1^{o} - X\hat{\Pi}^{o} = M_1(Y_1^{o} - X_1^{*}\hat{\Pi}_{2\cdot}^{o}) \tag{8}$$

Therefore,

$$\hat{W} = (Y_1^{o} - X\hat{\Pi}^{o})'(Y_1^{o} - X\hat{\Pi}^{o}) = (Y_1^{o} - X_1^{*}\hat{\Pi}_{2\cdot}^{o})'M_1(Y_1^{o} - X_1^{*}\hat{\Pi}_{2\cdot}^{o}) \tag{9}$$

But substituting the expression in (7) for $\Pi_{2\cdot}^{o}$ gives

$$\hat{W} = Y_1^{o\prime}M_1Y_1^{o} + [I - (\hat{\gamma}^{o\prime}\hat{W}\hat{\gamma}^{o})^{-1}\hat{W}\hat{\gamma}^{o}\hat{\gamma}^{o\prime}]Y_1^{o\prime}M_1X_1^{*}(X_1^{*\prime}M_1X_1^{*})^{-1}X_1^{*}$$
$$\cdot M_1X_1^{*}(X_1^{*\prime}M_1X_1^{*})^{-1}X_1^{*\prime}M_1Y_1^{o} \cdot [I - (\hat{\gamma}^{o\prime}\hat{W}\hat{\gamma}^{o})^{-1}\hat{\gamma}^{o}\hat{\gamma}^{o\prime}\hat{W}]$$
$$- [I - (\hat{\gamma}^{o\prime}\hat{W}\hat{\gamma}^{o})^{-1}\hat{W}\hat{\gamma}^{o}\hat{\gamma}^{o\prime}]Y_1^{o}M_1X_1^{*}(X_1^{*\prime}M_1X_1^{*})^{-1}X_1^{*\prime} \cdot M_1 \cdot Y_1^{o}$$
$$- Y_1^{o\prime}M_1X_1^{*}(X_1^{*\prime}M_1X_1^{*})^{-1}X_1^{*\prime}M_1Y_1^{o}[I - (\hat{\gamma}^{o\prime}\hat{W}\hat{\gamma}^{o})^{-1}\hat{\gamma}^{o}\hat{\gamma}^{o\prime}\hat{W}]$$

Simplifying this,

$$\hat{W} = Y_1^{o\prime}M_1Y_1^{o} - Y_1^{o}M_1X_1^{*}(X_1^{*\prime}M_1X_1^{*})^{-1}X_1^{*\prime}M_1Y_1^{o}$$
$$+ (\hat{\gamma}^{o\prime}\hat{W}\hat{\gamma}^{o})^{-2}\hat{W}\hat{\gamma}^{o}\hat{\gamma}^{o\prime}Y_1^{o\prime}M_1X_1^{*}(X_1^{*\prime}M_1X_1^{*})^{-1}X_1^{*\prime}M_1Y_1^{o}\hat{\gamma}^{o}\hat{\gamma}^{o\prime}\hat{W} \tag{10}$$

Using the fact that

$$M_1 - M = M_1X_1^{*}(X_1^{*\prime}M_1X_1^{*})^{-1}X_1^{*\prime}M_1 \qquad \text{(Lemma 8)}$$

where $M = I - X(X'X)^{-1}X'$, we have

$$\hat{W} = Y_1^{o}{}'M_1Y_1^{o} - Y_1^{o}{}'(M_1 - M)Y_1^{o} + (\hat{\gamma}^{o}{}'\hat{W}\hat{\gamma}^{o})^{-2}\hat{W}\hat{\gamma}^{o}\hat{\gamma}^{o}{}'Y_1^{o}{}'(M_1 - M)Y_1^{o}\hat{\gamma}^{o}\hat{\gamma}^{o}{}'\hat{W}$$

or

$$\hat{W} = Y_1^{o}{}'MY_1^{o} + \phi(\hat{\gamma}^{o}{}'\hat{W}\hat{\gamma}^{o})^{-1}\hat{W}\hat{\gamma}^{o}\hat{\gamma}^{o}{}'\hat{W} \tag{11}$$

where

$$\phi = (\hat{\gamma}^{o}{}'\hat{W}\hat{\gamma}^{o})^{-1}\hat{\gamma}^{o}{}'Y_1^{o}{}'(M_1 - M)Y_1^{o}\hat{\gamma}^{o} \tag{12}$$

Now, postmultiplying (11) by $\hat{\gamma}^{o}$ gives

$$\hat{W}\hat{\gamma}^{o} = Y_1^{o}{}'MY_1^{o}\hat{\gamma}^{o} + \phi(\hat{\gamma}^{o}{}'\hat{W}\hat{\gamma}^{o})^{-1}\hat{W}\hat{\gamma}^{o}\hat{\gamma}^{o}{}'\hat{W}\hat{\gamma}^{o}$$

$$= Y_1^{o}{}'MY_1^{o}\hat{\gamma}^{o} + \phi\hat{W}\hat{\gamma}^{o}$$

which implies

$$\hat{W}\hat{\gamma}^{o} = \frac{1}{1 - \phi} Y_1^{o}{}'MY_1^{o}\hat{\gamma}^{o} \tag{13}$$

Therefore

$$\hat{\gamma}^{o}{}'\hat{W}\hat{\gamma}^{o} = \frac{1}{1 - \phi} \hat{\gamma}^{o}{}'Y_1^{o}{}'MY_1^{o}\hat{\gamma}^{o} \tag{14}$$

We will now utilize the fact that $\hat{\Pi}_{2\cdot}^{o}{}'\lambda = 0$. To do so, substitute (7) for $\hat{\Pi}_{2\cdot}^{o}$ and (6) for λ to get

$$[I - (\hat{\gamma}^{o}{}'\hat{W}\hat{\gamma}^{o})^{-1}\hat{W}\hat{\gamma}^{o}\hat{\gamma}^{o}{}']Y_1^{o}{}'M_1X_1^{*}(X_1^{*}{}'M_1X_1^{*})^{-1}(\hat{\gamma}^{o}{}'\hat{W}\hat{\gamma}^{o})^{-1}X_1^{*}{}'M_1Y_1^{o}\hat{\gamma}^{o} = 0 \tag{15}$$

Multiplying by $\hat{\gamma}^{o}{}'\hat{W}\hat{\gamma}^{o}$ and using (13) and (14) to substitute out $\hat{W}\hat{\gamma}^{o}$ and $\hat{\gamma}^{o}{}'\hat{W}\hat{\gamma}^{o}$, we get

$$0 = Y_1^{o}{}'M_1X_1^{*}(X_1^{*}{}'M_1X_1^{*})^{-1}X_1^{*}{}'M_1Y_1^{o}\hat{\gamma}^{o}$$
$$- \frac{1 - \phi}{\hat{\gamma}^{o}{}'Y_1^{o}{}'MY_1^{o}\hat{\gamma}^{o}} \cdot \frac{1}{1 - \phi} Y_1^{o}{}'MY_1^{o}\hat{\gamma}^{o}\hat{\gamma}^{o}{}'Y_1^{o}{}'M_1X_1^{*}(X_1^{*}{}'M_1X_1^{*})^{-1}X_1^{*}{}'M_1Y_1^{o}\hat{\gamma}^{o}$$

Again using Lemma 8,

$$Y_1^{o\prime}(M_1 - M)Y_1^o\hat{\gamma}^o = Y_1^{o\prime}MY_1^o\hat{\gamma}^o \frac{\hat{\gamma}^{o\prime}Y_1^{o\prime}(M_1 - M)Y_1^o\hat{\gamma}^o}{\hat{\gamma}^{o\prime}Y_1^{o\prime}MY_1^o\hat{\gamma}^o}$$

Letting

$$\hat{\ell} = \frac{\hat{\gamma}^{o\prime}Y_1^{o\prime}M_1Y_1^o\hat{\gamma}^o}{\hat{\gamma}^{o\prime}Y_1^{o\prime}MY_1^o\hat{\gamma}^o}$$

this simplifies to

$$Y_1^{o\prime}M_1Y_1^o\hat{\gamma}^o - Y_1^{o\prime}MY_1^o\hat{\gamma}^o = Y_1^{o\prime}MY_1^o\hat{\gamma}^o(\hat{\ell} - 1)$$

or

$$(Y_1^{o\prime}M_1Y_1^o - \hat{\ell}Y_1^{o\prime}MY_1^o)\hat{\gamma}^o = 0 \tag{16}$$

But this is just the equation giving the LVR estimator; $\hat{\ell}$ is of course a root of

$$|Y_1^{o\prime}M_1Y_1^o - \hat{\ell}Y_1^{o\prime}MY_1^o| = 0$$

Indeed all that remains to be shown is that it is the *smallest* root of this equation which leads to the minimum of the generalized residual variance.

To show this, go back to equation (15), use (13) to substitute out $\hat{W}\hat{\gamma}^o$, but do not substitute out $\hat{\gamma}^{o\prime}\hat{W}\hat{\gamma}^o$; this yields

$$0 = [I - (\hat{\gamma}^{o\prime}\hat{W}\hat{\gamma}^o)^{-1}\frac{1}{1 - \phi}Y_1^{o\prime}MY_1^o\hat{\gamma}^o\hat{\gamma}^{o\prime}]Y_1^{o\prime}M_1X_1^*(X_1^{*\prime}M_1X_1^*)^{-1}X_1^{*\prime}M_1Y_1^o\hat{\gamma}^o$$

or (using Lemma 8)

$$Y_1^{o\prime}(M_1 - M)Y_1^o\hat{\gamma}^o = (\hat{\gamma}^{o\prime}\hat{W}\hat{\gamma}^o)^{-1}\frac{1}{1 - \phi}Y_1^{o\prime}MY_1^o\hat{\gamma}^o\hat{\gamma}^{o\prime}Y_1^{o\prime}(M_1 - M)Y_1^o\hat{\gamma}^o$$

Using (12), this simplifies to

$$Y_1^{o\prime}(M_1 - M)Y_1^o\hat{\gamma}^o = \frac{\phi}{1 - \phi}Y_1^{o\prime}MY_1^o\hat{\gamma}^o$$

or

$$(Y_1^{o\prime} M_1 Y_1^o - \hat{\ell} Y_1^{o\prime} M Y_1^o)\hat{\gamma}^o = 0 \tag{17}$$

as before, where now we have also the fact that

$$\hat{\ell} = 1 - \frac{\phi}{1 - \phi} = \frac{1}{1 - \phi} \tag{18}$$

As a matter of notation, let

$$\hat{\ell}_1, \hat{\ell}_2, \cdots, \hat{\ell}_{g_*}$$

be the roots of the equation $|Y_1^{o\prime} M_1 Y_1^o - \hat{\ell} Y_1^{o\prime} M Y_1^o| = 0$, and let $\hat{\gamma}_1^o, \hat{\gamma}_2^o, \cdots, \hat{\gamma}_{g_*}^o$ be the corresponding solutions to (17). Let

$$G = (\hat{\gamma}_1^o, \cdots, \hat{\gamma}_{g_*}^o)$$

be normalized so that

$$G'(Y_1^{o\prime} M Y_1^o)G = I. \tag{19}$$

Now if (13) is substituted in (11), we get

$$\hat{W} = Y_1^{o\prime} M Y_1^o + \frac{\phi_i}{1 - \phi_i} \frac{Y_1^{o\prime} M Y_1^o \hat{\gamma}_i^o \hat{\gamma}_i^{o\prime} Y_1^{o\prime} M Y_1^o}{\hat{\gamma}_i^{o\prime} Y_1^{o\prime} M Y_1^o \hat{\gamma}_i^o}$$

Since $\phi_i/(1 - \phi_i) = \hat{\ell}_i - 1$ and since $\hat{\gamma}_i^{o\prime} Y_1^{o\prime} M Y_1^o \hat{\gamma}_i^o = 1$ in light of (19), this simplifies to

$$\hat{W} = Y_1^{o\prime} M Y_1^o + (\hat{\ell}_i - 1) Y_1^{o\prime} M Y_1^o \hat{\gamma}_i^o \hat{\gamma}_i^{o\prime} Y_1^{o\prime} M Y_1^o$$

Pre- and postmultiplying by G' and G,

$$G'\hat{W}G = I + (\hat{\ell}_i - 1)e_i e_i'$$

where e_i is the i-th column of I. But $I + (\hat{\ell}_i - 1)e_i e_i'$ is an identity matrix except for $\hat{\ell}_i$ in the i-th position, and so

$$|G'\hat{W}G| = \hat{\ell}_i$$

Hence

$$|\hat{W}| = \hat{\ell}_1 |G|^{-2}$$

Since $|G|$ is fixed, we minimize $|\hat{W}|$ by choosing the smallest root $\hat{\ell}_1$.

This completes the proof of the theorem. However, it may be worth noting that since $G'(Y_1^{o\prime} M Y_1^o)G = I$,

$$|G|^{-2} = |Y_1^{o\prime} M Y_1^o|$$

Therefore

$$\hat{\ell} = \frac{|\hat{W}|}{|Y_1^{o\prime} M Y_1^o|}$$

where $|\hat{W}|$ is the minimized value of $|W|$.

13. COMMENT: If the equation is exactly identified, the constraint that rank $(\hat{\Pi}_{2\cdot}^o) = g_* - 1$ is no constraint at all. Hence we have an unconstrained minimization of $|\hat{W}|$; the solution is $\hat{W} = Y_1^{o\prime} M Y_1^o$ (as given by OLS). Therefore

$$\hat{\ell} = \frac{|\hat{W}|}{|Y_1^{o\prime} M Y_1^o|} = 1$$

This is of course an alternative proof of Proposition 9.

Limited Information Maximum Likelihood Estimation

Again return to the equation

$$Y_1^o \gamma^o + X_1 \delta + \varepsilon_{\cdot 1} = 0$$

and note that we can estimate γ^o by solving $\hat{\Pi}_{2\cdot}^o \hat{\gamma}^o = 0$, *given* that we have an estimate $\hat{\Pi}_{\cdot 2}^o$ whose rank is only $g_* - 1$. In the previous subsection we derived such an estimate by minimizing the generalized residual variance subject to the constraint that $\hat{\Pi}_{2\cdot}^o \hat{\gamma}^o = 0$. In this subsection we will maximize the joint likelihood function of Y_1^o subject to this constraint. The resulting estimator turns out to be identical to the LGVR estimator.

14. DEFINITION: The *limited information maximum likelihood* (LIML) estimator of γ^o is the estimator gotten by maximizing the joint likelihood function of Y_1^o subject to the constraint that $\Pi_{2.}^o\gamma^o = 0$.

15. LEMMA: Let A and B be square n × n matrices. Then

$$\frac{\partial\ \text{trace}(AB)}{\partial A} = B'$$

Proof:

$$\text{trace}(AB) = \sum_{i=1}^{n}\sum_{j=1}^{n} A_{ij}B_{ji}$$

Therefore

$$\frac{\partial\ \text{trace}(AB)}{\partial A_{km}} = B_{mk}; \qquad \frac{\partial\ \text{trace}(AB)}{\partial A} = B'$$

16. THEOREM: The LIML estimator is *identical* to the LGRV estimator (and the LVR estimator).

Proof: From Theorem 2 of Section 4.4, the likelihood function of the model

$$Y = X\Pi + V$$

is

$$L(Y) = (2\pi)^{-GT/2}|\Omega|^{-T/2}\exp[-\tfrac{1}{2}\ \text{trace}\ \Omega^{-1}(Y - X\Pi)'(Y - X\Pi)]$$

If we consider only Y_1^o, and look at the log likelihood function, we have

$$\log L(Y_1^o) = C_1 + \frac{T}{2}\log|\Omega_o^{-1}| - \tfrac{1}{2}\ \text{trace}\ \Omega_o^{-1}(Y_1^o - X\Pi^o)'(Y_1^o - X\Pi^o)$$

We now proceed to maximize this partially with respect to Ω_o^{-1}. This is permissible since the constraint $\Pi_{2.}^o\gamma^o$ has no effect on the maximization with respect to Ω_o^{-1}; formally, if λ is the vector of Lagrangian multipliers, then

$$\frac{\partial\lambda'\Pi_{2.}^o\gamma^o}{\partial\Omega_o^{-1}} = 0$$

Now, using Lemma 15 and the fact that

$$\frac{\partial \log|A|}{\partial A} = A^{-1}$$

we obtain

$$\frac{\partial \log L(Y_1^o)}{\partial \Omega_o^{-1}} = \frac{T}{2}\Omega_o - \frac{1}{2}(Y_1^o - X\Pi^o)'(Y_1^o - X\Pi^o) = 0$$

This yields

$$\Omega_o = \frac{1}{T}(Y_1^o - X\Pi^o)'(Y_1^o - X\Pi^o)$$

as a relationship which must hold at the maximum. We now make this substitution for Ω_o in the log likelihood function to obtain the "concentrated" log likelihood function

$$\begin{aligned}\log L^*(Y_1^o) &= C_1 - \frac{T}{2}\log\left|\frac{1}{T}(Y_1^o - X\Pi^o)'(Y_1^o - X\Pi^o)\right| - \frac{1}{2}\text{trace } \Omega_o^{-1}\Omega_o \\ &= C_2 - \frac{T}{2}\log\left|\frac{1}{T}(Y_1^o - X\Pi^o)'(Y_1^o - X\Pi^o)\right|\end{aligned}$$

This must now be maximized (with respect to Π^o) subject to the restriction that $\Pi_{2.}^o\gamma^o = 0$. But clearly this maximization is equivalent to the minimization of the generalized residual variance

$$|(Y_1^o - X\Pi^o)'(Y_1^o - X\Pi^o)|$$

subject to the same constraint, which defines the LGRV estimator.

The following derivation of the LIML estimator perhaps serves the purpose of making clear the sense in which LIML is a "limited information" estimator. We will proceed to directly maximize the likelihood function (expressed in terms of *structural* parameters), but ignoring all prior restrictions except those on the first equation.

First we will prove a lemma.

17. LEMMA: Let β be an $n \times G$ matrix and Σ be a $G \times G$ positive definite matrix. Then there exists a nonsingular $G \times G$ transformation matrix H such that

$$\beta H_{\cdot 1} = \beta_{\cdot 1}$$

$$H'\Sigma H = \begin{bmatrix} \sigma_{11} & 0 \\ 0 & I_{G-1} \end{bmatrix}$$

where $\beta_{\cdot 1}$ and $H_{\cdot 1}$ are the first columns of β and H, respectively, and where σ_{11} is the 1,1-th element of Σ.

Proof: Partition H as

$$H = \begin{bmatrix} H_{11} & H_{12} \\ H_{21} & H_{22} \end{bmatrix}$$

where H_{11} is a scalar and $H_{\cdot 1} = \begin{bmatrix} H_{11} \\ H_{21} \end{bmatrix}$. Then

$$\beta H_{\cdot 1} = [\beta_{\cdot 1}, \beta_{(1)}]H_{\cdot 1} = \beta_{\cdot 1}H_{11} + \beta_{(1)}H_{21}$$

Picking $H_{11} = 1$ and $H_{21} = 0$, the requirement that $\beta H_{\cdot 1} = \beta_{\cdot 1}$ is fulfilled.

Now, letting

$$H = \begin{bmatrix} 1 & H_{12} \\ 0 & H_{22} \end{bmatrix} \qquad \Sigma = \begin{bmatrix} \sigma_{11} & \Sigma_{12} \\ \Sigma_{21} & \Sigma_{22} \end{bmatrix}$$

we have

$$H'\Sigma H = \begin{bmatrix} \sigma_{11} & \sigma_{11}H_{12} + \Sigma_{12}H_{22} \\ H_{12}'\sigma_{11} + H_{22}'\Sigma_{21} & H_{12}'\sigma_{11}H_{12} + H_{22}'\Sigma_{21}H_{12} + H_{12}'\Sigma_{12}H_{22} + H_{22}'\Sigma_{22}H_{22} \end{bmatrix}$$

If we set

$$H_{12} = -\frac{\Sigma_{12}H_{22}}{\sigma_{11}}$$

then

$$H'\Sigma H = \begin{bmatrix} \sigma_{11} & 0 \\ 0 & -\dfrac{H_{22}'\Sigma_{21}}{\sigma_{11}}\left(-\sigma_{11}\dfrac{\Sigma_{12}H_{22}}{\sigma_{11}} + \Sigma_{12}H_{22}\right) + H_{22}'\left(-\dfrac{\Sigma_{21}\Sigma_{12}}{\sigma_{11}} + \Sigma_{22}\right)H_{22} \end{bmatrix}$$

$$= \begin{bmatrix} \sigma_{11} & 0 \\ 0 & H_{22}'\left(\Sigma_{22} - \dfrac{\Sigma_{21}\Sigma_{12}}{\sigma_{11}}\right)H_{22} \end{bmatrix}$$

Now, since Σ is positive definite, so is

$$\left(\Sigma_{22} - \frac{\Sigma_{21}\Sigma_{12}}{\sigma_{11}}\right) \qquad \text{[see Note 2]}$$

Hence there exists a nonsingular matrix A such that

$$A'A = \Sigma_{22} - \frac{\Sigma_{21}\Sigma_{12}}{\sigma_{11}}$$

But then

$$(A^{-1})'\left(\Sigma_{22} - \frac{\Sigma_{21}\Sigma_{12}}{\sigma_{11}}\right)A^{-1} = I$$

Picking $H_{22} = A^{-1}$, we do indeed have

$$H\Sigma H = \begin{bmatrix} \sigma_{11} & 0 \\ 0 & I \end{bmatrix}$$

Finally, then, the nonsingularity of

$$\begin{bmatrix} \sigma_{11} & 0 \\ 0 & I \end{bmatrix}$$

implies the nonsingularity of H.

Now let us return to the system

$$Y\Gamma + X\Delta + \varepsilon = 0$$

or

$$Z\beta + \varepsilon = 0 \qquad Z = (Y,X) \qquad \beta = \begin{pmatrix} \Gamma \\ \Delta \end{pmatrix}$$

As a matter of notation, let $\Gamma_{\cdot 1}$, $\Delta_{\cdot 1}$, and $\beta_{\cdot 1}$ be the first columns of Γ, Δ, and β, respectively, and let

$$\Gamma = [\Gamma_{\cdot 1}, \Gamma_{(1)}] \qquad \Delta = [\Delta_{\cdot 1}, \Delta_{(1)}] \qquad \beta = [\beta_{\cdot 1}, \beta_{(1)}]$$

Now let us transform the system by a nonsingular matrix H in such a way that

$$\beta H = [\beta_{\cdot 1}, \beta_{(1)}]H = [\beta_{\cdot 1}, B_*] \quad , \qquad B_* = \begin{pmatrix} \Gamma_* \\ \Delta_* \end{pmatrix}$$

$$H'\Sigma H = \begin{pmatrix} \sigma_{11} & 0 \\ 0 & I \end{pmatrix}$$

(This is possible by Lemma 17.) Note that the transformation leaves the parameters of the first equation unchanged, and that the first equation is now independent of the last G - 1 transformed equations.

18. THEOREM: If the likelihood function of the above system is maximized subject only to the constraints on the first equation, the resulting estimator is the LIML estimator.

Proof: From Theorem 2 of Section 4.4, the log likelihood function of the system is

$$L = C_1 - \frac{T}{2}\log|\Sigma| + T\log||\Gamma|| - \frac{1}{2}\operatorname{trace}\Sigma^{-1}(Z\beta)'(Z\beta)$$

Since the transformation by the nonsingular matrix H does not affect the value of the likelihood function, we can write it also in terms of the transformed system as

$$L = C_1 - \frac{T}{2}\log|H'\Sigma H| + T\log||\Gamma H|| - \frac{1}{2}\operatorname{trace}(H'\Sigma H)^{-1}(Z\beta H)'(Z\beta H)$$

Using the facts that

$$\beta H = [\beta_{\cdot 1}, B_*] = \begin{bmatrix} \Gamma_{\cdot 1} & \Gamma_* \\ \Delta_{\cdot 1} & \Delta_* \end{bmatrix}$$

$$H'\Sigma H = \begin{bmatrix} \sigma_{11} & 0 \\ 0 & I \end{bmatrix}$$

we have

$$L = C_1 - \frac{T}{2} \log \begin{bmatrix} \sigma_{11} & 0 \\ 0 & I \end{bmatrix} + T \log||\Gamma_{\cdot 1}, \Gamma_*||$$

$$- \frac{1}{2} \operatorname{trace} \begin{bmatrix} \sigma_{11}^{-1} & 0 \\ 0 & I \end{bmatrix} (\beta_{\cdot 1}, B_*)'Z'Z(\beta_{\cdot 1}, B_*)$$

Therefore

$$L = C_1 - \frac{T}{2} \log \sigma_{11} + T \log||\Gamma_{\cdot 1}, \Gamma_*|| - \frac{1}{2\sigma_{11}} \beta_{\cdot 1}'Z'Z'\beta_{\cdot 1}$$

$$- \frac{1}{2} \operatorname{trace} B_*'Z'ZB_* \qquad (1)$$

We now proceed to eliminate B_* by maximizing partially with respect to it, *ignoring all restrictions on* B_*.

$$\frac{\partial L}{\partial B_*} = T \frac{\partial \log||\Gamma_{\cdot 1}, \Gamma_*||}{\partial B_*} - \frac{1}{2} \frac{\partial \operatorname{trace} B_*'Z'ZB_*}{\partial B_*} = 0 \qquad (2)$$

Now

$$\frac{\partial \log||\Gamma_{\cdot 1}, \Gamma_*||}{\partial B_*} = \begin{bmatrix} \dfrac{\partial \log||\Gamma_{\cdot 1}, \Gamma_*||}{\partial \Gamma_*} \\ \\ \dfrac{\partial \log||\Gamma_{\cdot 1}, \Gamma_*||}{\partial \Delta_*} \end{bmatrix}$$

Furthermore

$$\frac{\partial \log||\Gamma_{\cdot 1}, \Gamma_*||}{\partial \Delta_*} = 0$$

and

$$\frac{\partial \log||\Gamma_{\cdot 1}, \Gamma_*||}{\partial \Gamma_*} = J$$

where J consists of the last G - 1 columns of $(\Gamma_{\cdot 1}, \Gamma_*)'^{-1}$; that is,

$$J = \begin{pmatrix} \Gamma'_{\cdot 1} \\ \Gamma'_* \end{pmatrix}^{-1} \begin{pmatrix} 0 \\ I_{G-1} \end{pmatrix} \tag{3}$$

Also

$$\frac{\partial \text{ trace } B'_* Z'ZB_*}{\partial B_*} = 2Z'ZB_*$$

Hence, with these substitutions into (2) we have

$$\frac{\partial L}{\partial B_*} = T\begin{pmatrix} J \\ 0 \end{pmatrix} - Z'ZB_* = 0$$

or

$$Z'ZB_* = \begin{pmatrix} T\ J \\ 0 \end{pmatrix} \tag{4}$$

Therefore

$$\begin{aligned} B'_*Z'ZB_* &= [\Gamma'_*, \Delta'_*]\begin{pmatrix} T\ J \\ 0 \end{pmatrix} \\ &= T\Gamma'_*J \\ &= T[0,I]\begin{pmatrix} \Gamma'_{\cdot 1} \\ \Gamma'_* \end{pmatrix}\begin{pmatrix} \Gamma'_{\cdot 1} \\ \Gamma'_* \end{pmatrix}^{-1}\begin{pmatrix} 0 \\ I \end{pmatrix} \qquad \text{[using (3)]} \end{aligned}$$

Hence at the maximum we have

$$-\frac{1}{2} \text{ trace } B'_*Z'ZB_* = -\frac{T}{2}(G - 1) \tag{5}$$

which is a constant.

We also need the maximized value of $T \log||\Gamma_{\cdot 1}, \Gamma_*||$. Now

$$\begin{pmatrix} TJ \\ 0 \end{pmatrix} = Z'ZB_* = \begin{pmatrix} Y'Y & Y'X \\ X'Y & X'X \end{pmatrix} \begin{pmatrix} \Gamma_* \\ \Delta_* \end{pmatrix}$$

so that

$$\begin{aligned} TJ &= Y'Y\Gamma_* + Y'X\Delta_* \\ 0 &= X'Y\Gamma_* + X'X\Delta_* \end{aligned} \tag{6}$$

The second of these two equations yields

$$\Delta_* = -(X'X)^{-1}X'Y\Gamma_*$$

Substituting this in the first gives

$$TJ = [Y'Y - Y'X(X'X)^{-1}X'Y]\Gamma_* = Y'MY\Gamma_*$$

where $M - I - X(X'X)^{-1}X'$. This can be written as

$$J = W\Gamma_* \qquad W = \frac{1}{T} Y'MY \tag{7}$$

Now, it is an identity that

$$T \log||\Gamma_{\cdot 1},\Gamma_*|| = \frac{T}{2} \log[||\Gamma_{\cdot 1},\Gamma_*||^2|W|] - \frac{T}{2} \log|W| \tag{8}$$

Note that $|W|$ is a constant in the sense that it does not depend on any parameters. Also

$$||\Gamma_{\cdot 1},\Gamma_*||^2|W| = |(\Gamma_{\cdot 1},\Gamma_*)'W(\Gamma_{\cdot 1},\Gamma_*)|$$

$$\begin{vmatrix} \Gamma_{\cdot 1}'W\Gamma_{\cdot 1} & \Gamma_{\cdot 1}'W\Gamma_* \\ \Gamma_*'W\Gamma_{\cdot 1} & \Gamma_*'W\Gamma_* \end{vmatrix} \tag{9}$$

Since

$$\begin{aligned} W\Gamma_* &= J & \text{[by (7)]} \\ &= \begin{pmatrix} \Gamma_{\cdot 1}' \\ \Gamma_*' \end{pmatrix}^{-1} \begin{pmatrix} 0 \\ I \end{pmatrix} & \text{[by (3)]} \end{aligned}$$

we have

$$\Gamma_{\cdot 1}' W \Gamma_* = [I, 0] \begin{pmatrix} \Gamma_{\cdot 1}' \\ \Gamma_*' \end{pmatrix} \begin{pmatrix} \Gamma_{\cdot 1}' \\ \Gamma_*' \end{pmatrix}^{-1} \begin{pmatrix} 0 \\ I \end{pmatrix} = 0$$

and

$$\Gamma_*' W \Gamma_* = [0, I] \begin{pmatrix} \Gamma_{\cdot 1}' \\ \Gamma_*' \end{pmatrix} \begin{pmatrix} \Gamma_{\cdot 1}' \\ \Gamma_*' \end{pmatrix}^{-1} \begin{pmatrix} 0 \\ I \end{pmatrix} = I_{G-1}$$

Substituting back into (9) we get

$$||\Gamma_{\cdot 1}, \Gamma_*||^2 |W| = \begin{vmatrix} \Gamma_{\cdot 1}' W \Gamma_{\cdot 1} & 0 \\ 0 & I \end{vmatrix} = |\Gamma_{\cdot 1}' W \Gamma_{\cdot 1}|$$

Substituting this into (8) yields

$$T \log ||\Gamma_{\cdot 1}, \Gamma_*|| = \frac{T}{2} \log \Gamma_{\cdot 1}' W \Gamma_{\cdot 1} - \frac{T}{2} \log |W| \tag{10}$$

at the maximum point.

Now, substituting (10) and (5) back into the original likelihood function, we obtain the "concentrated" likelihood function

$$L^* = C_1 - \frac{T}{2} \log \sigma_{11} + \frac{T}{2} \log \Gamma_{\cdot 1}' W \Gamma_{\cdot 1} - \frac{T}{2} \log |W| - \frac{1}{2\sigma_{11}} \beta_{\cdot 1}' Z' Z \beta_{\cdot 1}$$

or

$$L^* = C_2 - \frac{T}{2} \log \sigma_{11} + \frac{T}{2} \log \Gamma_{\cdot 1}' W \Gamma_{\cdot 1} - \frac{1}{2\sigma_{11}} \beta_{\cdot 1}' Z' Z \beta_{\cdot 1} \tag{11}$$

where W is defined in (7). We now have the likelihood function expressed in terms of only the parameters of the first equation, and we wish to maximize it, imposing all prior constraints on the first equation. To do so, let

$$\Gamma_{\cdot 1} = \begin{pmatrix} \gamma^o \\ 0 \end{pmatrix} \qquad \Delta_{\cdot 1} = \begin{pmatrix} \delta \\ 0 \end{pmatrix} \qquad Y = (Y_1^o, Y_1^*) \qquad X_1 = (X_1, X_1^*) \tag{12}$$

We also partition W as

$$W = \begin{pmatrix} W_{11} & W_{12} \\ W_{21} & W_{22} \end{pmatrix} \tag{13}$$

so that

$$W_{11} = \frac{1}{T} Y_1^{o\prime} M Y_1^o \tag{14}$$

Also note that in this notation

$$\Gamma_{\cdot 1}' W \Gamma_{\cdot 1} = \begin{pmatrix} \gamma^o \\ 0 \end{pmatrix}' \begin{pmatrix} W_{11} & W_{12} \\ W_{21} & W_{22} \end{pmatrix} \begin{pmatrix} \gamma^o \\ 0 \end{pmatrix} = \gamma^{o\prime} W_{11} \gamma^o \tag{15}$$

Now, since $Z = [Y_1^o, Y_1^*, X_1, X_1^*]$

$$\beta_{\cdot 1}' Z' Z \beta_{\cdot 1} = \begin{pmatrix} \gamma^o \\ 0 \\ \delta \\ 0 \end{pmatrix}' \begin{pmatrix} Y_1^{o\prime} Y_1^o & Y_1^{o\prime} Y_1^* & Y_1^{o\prime} X_1 & Y_1^{o\prime} X_1^* \\ Y_1^{*\prime} Y_1^o & Y_1^{*\prime} Y_1^* & Y_1^{*\prime} X_1 & Y_1^{*\prime} X_1^* \\ X_1' Y_1^o & X_1' Y_1^* & X_1' X_1 & X_1' X_1^* \\ X_1^{*\prime} Y_1^o & X_1^{*\prime} Y_1^* & X_1^{*\prime} X_1 & X_1^{*\prime} X_1^* \end{pmatrix} \begin{pmatrix} \gamma^o \\ 0 \\ \delta \\ 0 \end{pmatrix}$$

or

$$\beta_{\cdot 1}' Z' Z \beta_{\cdot 1} = \gamma^{o\prime} Y_1^{o\prime} Y_1^o \gamma^o + 2\delta' X_1' Y_1^o \gamma^o + \delta' X_1' X_1 \delta \tag{16}$$

Substituting (16) and (15) back into (11), we get the concentrated likelihood function expressed as

$$L^* = C_2 - \frac{T}{2} \sigma_{11} + \frac{T}{2} \log \gamma^{o\prime} W_{11} \gamma^o - \frac{1}{2\sigma_{11}} (\gamma^{o\prime} Y_1^{o\prime} Y_1^o \gamma^o + 2\delta' X_1' Y_1^{o\prime} \gamma^o + \delta' X_1' X_1 \delta) \tag{17}$$

We now maximize partially with respect to δ:

$$\frac{\partial L^*}{\partial \delta} = - \frac{1}{2\sigma_{11}} (2X_1' Y_1^{o\prime} \gamma^o + 2X_1' X_1 \delta) = 0$$

which gives

$$\delta = - (X_1' X_1)^{-1} X_1' Y_1 \gamma^o \tag{18}$$

the usual result. Now, substituting this back into (17) gives

$$L^{**} = C_2 - \frac{T}{2}\log \sigma_{11} + \frac{T}{2}\log \gamma^{o\prime}W_{11}\gamma^o$$
$$- \frac{1}{2\sigma_{11}}[\gamma^{o\prime}Y_1^{o\prime}Y_1^o\gamma^o - 2\gamma^{o\prime}Y_1^{o\prime}X_1(X_1'X_1)^{-1}X_1'Y_1^o\gamma^o$$
$$+ \gamma^{o\prime}Y_1^{o\prime}X_1(X_1'X_1)^{-1}X_1'X_1(X_1'X_1)^{-1}X_1'Y_1^o\gamma^o]$$

or

$$L^{**} = C_2 - \frac{T}{2}\log \sigma_{11} + \frac{T}{2}\log \gamma^{o\prime}W_{11}\gamma^o - \frac{T}{2\sigma_{11}}\gamma^{o\prime}W_{11}^*\gamma^o \qquad (19)$$

where

$$W_{11}^* = \frac{1}{T}Y_1^{o\prime}[I - X_1(X_1'X_1)^{-1}X_1']Y_1^o = \frac{1}{T}Y_1^{o\prime}M_1Y_1^o \qquad (20)$$

Finally, we maximize (19) partially with respect to σ_{11}:

$$\frac{\partial L^{**}}{\partial \sigma_{11}} = -\frac{T}{2\sigma_{11}} + \frac{T}{2\sigma_{11}^2}\gamma^{o\prime}W_{11}^*\gamma^o = 0$$

which gives

$$\sigma_{11} = \gamma^{o\prime}W_{11}^*\gamma^o \qquad (21)$$

Inserting (21) in (19) gives

$$L^{***} = C_2 - \frac{T}{2}\log \gamma^{o\prime}W_{11}^*\gamma^o + \frac{T}{2}\log \gamma^{o\prime}W_{11}\gamma^o - \frac{T}{2}$$

$$L^{***} = C_3 - \frac{T}{2}\log\left(\frac{\gamma^{o\prime}W_{11}^*\gamma^o}{\gamma^{o\prime}W_{11}\gamma^o}\right) \qquad (22)$$

This shows that maximizing the likelihood function is equivalent to minimizing the ratio

$$\frac{\gamma^{o\prime}W_{11}^*\gamma^o}{\gamma^{o\prime}W_{11}\gamma^o}$$

with respect to γ^o. But this ratio is just

$$\ell = \frac{\gamma^{o\prime}Y_1^{o\prime}M_1Y_1^o\gamma^o}{\gamma^{o\prime}Y_1^{o\prime}MY_1^o\gamma^o}$$

and minimizing this with respect to γ^o defines the LVR estimator.

EXERCISES

1. Let $\hat{Y}_1 = X\hat{\Pi}_{\cdot 2}$, where $\hat{\Pi}_{\cdot 2}$ is any consistent estimator of $\Pi_{\cdot 2}$ (the portion of the reduced form corresponding to Y_1). Consider the estimator

$$\begin{pmatrix} \hat{\gamma} \\ \hat{\delta} \end{pmatrix} = [(\hat{Y}_1, X_1)'(Y_1, X_1)]^{-1}(\hat{Y}_1, X_1)'y$$

Show that it is consistent, and that its asymptotic distribution is identical to that of the 2SLS estimator.

2. Show that when the number of observations T is less than the number of predetermined variables, the 2SLS estimator does not exist.

3. Prove that the 2SLS residuals add to zero if the equation contains a constant term.

4. In the reduced form, the disturbances are correlated across equations. Explain why the seemingly unrelated regressions technique does *not* give more efficient estimates of the reduced-form parameters than OLS, despite this correlation across equations.

5. Are there any reasonable circumstances under which the condition of Proposition 32 of Section 4.4, that

$$\text{rank}\left\{\begin{pmatrix} \phi \\ N \end{pmatrix}(I \otimes B)\right\} = G^2$$

reduces to the condition that

$$\text{rank}[\phi(I \otimes B)] = G^2 - G \text{ ?}$$

NOTES

Section 4.3

1. If A is an $m \times n$ matrix with columns $A_{\cdot 1}, \ldots, A_{\cdot n}$, then

$$\text{vec } A = \begin{pmatrix} A_{\cdot 1} \\ A_{\cdot 2} \\ \vdots \\ A_{\cdot n} \end{pmatrix}$$

2. Used here is the fact that $\text{vec}(ABC) = [C' \otimes A]\text{vec}(B)$. See Nissen (1968).

Section 4.4

1. The fact used here is that for any matrix B, the rank of B plus the rank of the null space of B equals the number of columns of B.
2. Note that $A'\Sigma A$ is the covariance matrix of the transformed disturbance εA.

Section 4.8

1. Used here is the fact that

$$\frac{d}{dt}\frac{X}{Y} = \frac{1}{Y}\frac{dX}{dt} - \frac{X}{Y^2}\frac{dY}{dt} \qquad \text{(X and Y scalars)}$$

and that

$$\frac{d}{d\xi}\xi' A\xi = 2A\xi \qquad \text{(A a symmetric matrix and } \xi \text{ a vector)}$$

2. If Σ is the covariance matrix of the distribution of some n-dimensional random variable, then

$$\Sigma_{22} - \frac{\Sigma_{21}\Sigma_{22}}{\sigma_{11}}$$

is the covariance matrix of the distribution of the last n - 1 elements conditional on the first.

REFERENCES

Anderson, T.W., and H. Rubin (1949), "Estimation of the Parameters of a Single Equation in a Complete System of Stochastic Equations," *Annals of Mathematical Statistics 20*, 46-63.

Anderson, T.W., and H. Rubin (1950), "The Asymptotic Properties of Estimates of the Parameters of a Single Equation in a Complete System of Stochastic Equations," *Annals of Mathematical Statistics 21*, 570-582.

Basmann, R.L. (1957), "A Generalized Classical Method of Linear Estimation of Coefficients in a Structural Equation," *Econometrica 25*, 77-83.

Bowden, R. (1973), "The Theory of Parametric Identification," *Econometrica 41*, 1069-1074.

Brundy, J., and D.W. Jorgenson (1971), "Efficient Estimation of Simultaneous Equations by Instrumental Variables," *Review of Economics and Statistics 53*, 207-224.

Court, R.H. (1974), "Three Stage Least Squares and Some Extensions Where the Structural Disturbance Covariance Matrix May Be Singular," *Econometrica 42*, 547-559.

Dhrymes, P.J. (1970), *Econometrics: Statistical Foundations and Applications* New York: Harper and Row.

Fisher, F.M. (1966), *The Identification Problem in Econometrics* New York: McGraw-Hill.

Goldberger, A.S. (1964), *Econometric Theory* New York: John Wiley.

Goldberger, A.S. (1965), "An Instrumental Variables Interpretation of k-Class Estimation," *Indian Economic Journal 14*, 424-431.

Goldberger, A.S., and I. Olkin (1971), "A Minimum Distance Interpretation of Limited Information Estimation," *Econometrica 39*, 635-640.

Haavelmo, T. (1943), "The Statistical Implications of a System of Simultaneous Equations," *Econometrica 11*, 1-12.

Hood, W.C., and T.C. Koopmans (eds.) (1953), *Studies in Econometric Method* New York: John Wiley.

Kelly, J.S. (1975), "Linear Cross-Equation Constraints and the Identification Problem," *Econometrica 43*, 125-140.

Koopmans, T.C. (1945), "Statistical Estimation of Simultaneous Economic Relationships," *Journal of the American Statistical Association 40*, 448-466.

Koopmans, T.C., and O. Riersol (1950, "The Identification of Structural Characteristics," *Annals of Mathematical Statistics 21*, 165-181.

Maddala, G.S. (1971), "Simultaneous Estimation Methods for Large- and Medium-Size Econometric Models," *Review of Economic Studies 38*, 435-445.

Malinvaud, E. (1966), *Statistical Methods of Econometrics* Chicago: Rand McNally.

Mosback, E.J. and H.O. Wold (1970), *Interdependent Systems: Structure and Estimation* Amsterdam: North Holland.

Nissen, D. (1968), "A Note on the Variance of a Matrix," *Econometrica 36*, 603-604.

Rothenberg, T.J. (1971), "Identification in Parametric Models," *Econometrica 39*, 577-592.

Theil, H. (1961), *Economic Forecasts and Policy* Amsterdam: North Holland.

Theil, H. (1971), *Principles of Econometrics* New York: John Wiley.

Wegge, L. (1965), "Identifiability Criteria for a System of Equations as a Whole," *Australian Journal of Statistics 7*, 67-77.

CHAPTER 5

SIMULTANEOUS EQUATIONS II

5.1 INTRODUCTION

The simultaneous equations model to be considered in this chapter is the same as in Chapter 4. However, Chapter 4 discussed "single-equation" methods of estimation, in the sense that the estimators there operated on each equation separately. This chapter will discuss "systems" methods of estimation, which estimate all equations jointly.

The motivation for considering joint estimation procedures is of course that they are generally more (asymptotically) efficient than the single-equation procedures.

5.2 THREE-STAGE LEAST SQUARES

Again, let us consider the simultaneous equations model $Y\Gamma + X\Delta + \varepsilon = 0$. Eliminating the variables whose coefficients are known a priori to equal zero, we have the system of equations

$$y_i = Y_i\gamma_i + X_i\delta_i + \varepsilon_{\cdot i} = Z_i\beta_i + \varepsilon_{\cdot i} \qquad (i = 1, 2, \cdots, G)$$

where of course

$$Z_i = (Y_i, X_i) \qquad \beta_i = \begin{pmatrix} \gamma_i \\ \delta_i \end{pmatrix}$$

We now write this set of equations as

$$y_* = Z_*\alpha + \varepsilon_*$$

where

$$y_* = \begin{pmatrix} y_1 \\ y_2 \\ \vdots \\ y_G \end{pmatrix}$$

$$Z_* = \begin{pmatrix} Z_1 & & & 0 \\ & Z_2 & & \\ & & \ddots & \\ 0 & & & Z_G \end{pmatrix}$$

$$\alpha = \begin{pmatrix} \beta_1 \\ \beta_2 \\ \vdots \\ \beta_G \end{pmatrix}$$

$$\varepsilon_* = \text{vec}\ \varepsilon = \begin{pmatrix} \varepsilon_{\cdot 1} \\ \varepsilon_{\cdot 2} \\ \vdots \\ \varepsilon_{\cdot G} \end{pmatrix}$$

Note that the covariance matrix of ε_* is $\Sigma \otimes I_T$.

Now, the 2SLS estimator is just the OLS estimator in the set of equations

$$y_* = \hat{Z}_*\alpha + \varepsilon_*$$

which differs from the above system in that Z_* has been replaced by

$$\hat{Z}_* = \begin{pmatrix} \hat{Z}_1 & & & 0 \\ & \hat{Z}_2 & & \\ & & \ddots & \\ 0 & & & \hat{Z}_G \end{pmatrix}$$

where $\hat{Z}_i = X(X'X)^{-1}X'Z_i = [X(X'X)^{-1}X'Y_i, X_i] = (\hat{Y}_i, X_i)$. Thus the 2SLS estimator takes account of the fact that Z_* contains endogenous variables; however, it does *not* take account of the fact that the covariance matrix of ε_* is $\Sigma \otimes I_T$ rather than $\sigma^2 I_{GT}$.

To take account of the fact that there *is* correlation across equations, generalized least squares ought to be applied. To do so, given that Σ is unknown, requires an estimate of Σ, which can be gotten from the 2SLS residuals. This is the motivation for the following procedure.

1. DEFINITION: Consider the system of equations

$$y_* = Z_*\alpha + \varepsilon_*$$

as above. Let Σ be estimated by S, defined by

$$S_{ij} = \frac{1}{T}(y_i - Z_i\hat{\beta}_i)'(y_j - Z_j\hat{\beta}_j) \qquad (i, j = 1, 2, \cdots, G)$$

the $\hat{\beta}_i$ being the 2SLS estimates. Then the *three-stage least squares* (3SLS) estimator of α is

$$\hat{\alpha} = [\hat{Z}_*'(S^{-1} \otimes I)\hat{Z}_*]^{-1}\hat{Z}_*'(S^{-1} \otimes I)y_*$$

2. CAUTION: This does *not* amount to generalized least squares applied to the modified model $y_* = \hat{Z}_*\alpha + \varepsilon$. The reason is that the 2SLS residuals must be calculated as $y_i - Z_i\hat{\beta}_i$, not as $y_i - \hat{Z}_i\hat{\beta}_i$.

3. NOTE: If we denote the i,j-th element of S^{-1} by S^{ij}, then the i,j-th block of $\hat{Z}_*'(S^{-1} \otimes I)\hat{Z}_*$ is just

$$S^{ij}\hat{Z}_i'\hat{Z}_j = S^{ij}\begin{pmatrix} \hat{Y}_i'Y_j & Y_i'X_j \\ X_i'Y_j & X_i'X_j \end{pmatrix}$$

The i-th subvector of $\hat{Z}_*'(S^{-1} \otimes I)y_*$ is

$$\sum_{j=1}^{G} s^{ij}\hat{Z}_i'y_j = \sum_{j=1}^{G} s^{ij}\begin{pmatrix} \hat{Y}_i'y_j \\ X_i'y_j \end{pmatrix}$$

An alternative derivation of 3SLS proceeds as follows. We can premultiply the model $y_* = Z_*\alpha + \varepsilon_*$ by $(I_G \otimes X')$, which yields

$$(I \otimes X')y_* = (I \otimes X')Z_*\alpha + (I \otimes X')\varepsilon_*$$

(Here X is the matrix of observations on all predetermined variables.) This amounts to

$$\begin{pmatrix} X'y_1 \\ \vdots \\ X'y_G \end{pmatrix} = \begin{pmatrix} X'Z_1 & & 0 \\ & \ddots & \\ 0 & & X'Z_G \end{pmatrix}\alpha + \begin{pmatrix} X'\varepsilon_{\cdot 1} \\ \vdots \\ X'\varepsilon_{\cdot G} \end{pmatrix}$$

so that what has been done is to premultiply each equation by X'.

Now, 2SLS amounts to applying GLS to this system, using $I_G \otimes (X'X)$ as the covariance matrix of $(I \otimes X')\varepsilon_*$. On the other hand, using $\Sigma \otimes (X'X)$ as the covariance matrix of $(I \otimes X')\varepsilon_*$ gives

$$\hat{\alpha} = \{Z_*'(I \otimes X')'[\Sigma^{-1} \otimes (X'X)^{-1}](I \otimes X')Z_*\}^{-1}$$
$$\cdot\, Z_*'(I \otimes X')'[\Sigma^{-1} \otimes (X'X)^{-1}](I \otimes X')y_*$$

where we have used the fact that $(A \otimes B)^{-1} = A^{-1} \otimes B^{-1}$. Now, using the fact that $(A \otimes B)(C \otimes D) = AC \otimes BD$, we get

$$\hat{\alpha} = \{Z_*'[\Sigma^{-1} \otimes X(X'X)^{-1}X']Z_*\}^{-1}Z_*'[\Sigma^{-1} \otimes X(X'X)^{-1}X']y_*$$

As the following proposition indicates, this is the 3SLS estimator if Σ is replaced by its estimate S.

4. PROPOSITION: Let S be as defined in Definition 1. Then the estimator

$$\hat{\alpha} = \{Z_*'[S^{-1} \otimes X(X'X)^{-1}X']Z_*\}^{-1}Z_*'[S^{-1} \otimes X(X'X)^{-1}X']y_*$$

is the 3SLS estimator.

Proof: The i,j-th block of $Z_*'(S^{-1} \otimes X(X'X)^{-1}X')Z_*$ is

$$s^{ij}Z_i'X(X'X)^{-1}X'Z_j = s^{ij}\begin{pmatrix} \hat{Y}_i'Y_j & Y_i'X_j \\ X_i'Y_j & X_i'X_j \end{pmatrix}$$

which is identical to the i,j-th block of $\hat{Z}_*'(S^{-1} \otimes I)\hat{Z}_*$. Also, the i-th subvector of $Z_*'(S^{-1} \otimes X(X'X)^{-1}X')y_*$ is

$$\sum_{j=1}^{G} s^{ij}Z_i'X(X'X)^{-1}X'y_j = \sum_{j=1}^{G} s^{ij}\begin{pmatrix} \hat{Y}_i'y_j \\ X_i'y_j \end{pmatrix}$$

which is identical to the i-th subvector of $\hat{Z}_*'(S^{-1} \otimes I)y_*$.

Asymptotic Properties of the 3SLS Estimator

5. LEMMA: $\hat{\alpha} - \alpha = H(I \otimes X')\epsilon_*$, where

$$H = \{Z_*'[S^{-1} \otimes X(X'X)^{-1}X']Z_*\}^{-1}Z_*'(I \otimes X)[S^{-1} \otimes (X'X)^{-1}]$$

6. LEMMA: If all equations are identified by exclusion restrictions, and if Σ is nonsingular, then

$$\text{plim}\ \frac{1}{T}\{Z_*'[S^{-1} \otimes X(X'X)^{-1}X']Z_*\}$$

is nonsingular.

Proof: The i,j-th block of the above matrix is equal to

$$\sigma^{ij}\ \text{plim}\ \frac{1}{T}\begin{pmatrix} \hat{Y}_i'Y_j & Y_i'X_j \\ X_i'Y_j & X_i'X_j \end{pmatrix}$$

Now, as a matter of notation (which necessarily grows cumbersome), let

$$\underset{\sim}{Q} = \text{plim}\ \frac{1}{T}X'X$$

$$\underset{\sim}{Q}_i = \text{plim}\ \frac{1}{T}X'X_i$$

$$\underset{\sim}{Q}_{ij} = \text{plim}\ \frac{1}{T}X_i'X_j$$

and let Π_i be the submatrix of the reduced-form coefficient matrix

Π which corresponds to Y_i. Then

$$\text{plim}\,\frac{1}{T}\,\hat{Y}_i'Y_j = \Pi_i'\underset{\sim}{Q}\,\underset{\sim}{Q}^{-1}\underset{\sim}{Q}\Pi_j = \Pi_i'\underset{\sim}{Q}\Pi_j$$

$$\text{plim}\,\frac{1}{T}\,Y_i'X_j = \Pi_i'\underset{\sim}{Q}_j$$

$$\text{plim}\,\frac{1}{T}\,X_i'Y_j = \underset{\sim}{Q}_i'\Pi_j$$

$$\text{plim}\,\frac{1}{T}\,X_i'X_j = \underset{\sim}{Q}_{ij}$$

Hence the i,j-th block can be written as

$$\sigma^{ij}\begin{pmatrix} \Pi_i'\underset{\sim}{Q}\Pi_j & \Pi_i'\underset{\sim}{Q}_j \\ Q_i'\Pi_j & \underset{\sim}{Q}_{ij} \end{pmatrix}$$

Now, we can write $\underset{\sim}{Q}_i = \underset{\sim}{Q}C_i$, where C_i is the matrix, such that $X_i = XC_i$; that is, the column dimension of C_i is the number of predetermined variables in the i-th equation, and each is a *different* column of the K-dimensional identity matrix. (Note that therefore C_i has full column rank.) Also, we can write $\underset{\sim}{Q}_{ij} = C_i'\underset{\sim}{Q}C_j$. With these substitutions, the above block becomes

$$\sigma^{ij}\begin{pmatrix} \Pi_i'\underset{\sim}{Q}\Pi_j & \Pi_i'\underset{\sim}{Q}C_j \\ C_i'\underset{\sim}{Q}\Pi_j & C_i'\underset{\sim}{Q}C_j \end{pmatrix}$$

Also, since $\underset{\sim}{Q}$ is positive definite, there exists a *nonsingular* matrix V such that $\underset{\sim}{Q} = V'V$. With this substitution we get

$$\sigma^{ij}\begin{pmatrix} \Pi_i'V'V\Pi_j & \Pi_i'V'V'C_j \\ C_i'V'V\Pi_j & C_i'V'VC_j \end{pmatrix} = \sigma^{ij}(V\Pi_i, VC_i)'(V\Pi_j, VC_j)$$

Now, the original matrix under consideration, whose i,j-th block is the above, can hence be written as

$$P'(\Sigma^{-1} \otimes I)P$$

where

$$P = \text{diagonal}(V\Pi_i, VC_i)$$

Given that Σ is nonsingular, this will be nonsingular if and only if P is of full column rank, and this in turn will be true if and only if each diagonal block of P is of full column rank; that is, if

$$(V\Pi_i, VC_i) \qquad (i = 1, 2, \cdots, G)$$

is of full column rank.

Since V is nonsingular, this will be true if and only if (Π_i, C_i) is of full column rank for all i. And this, in turn, is guaranteed if the rank condition for identification holds (for each equation), as is shown in Lemma 5 of Section 4.6. To see this, note that the matrix (Π_i, C_i) can, with rearrangement of rows, be put into the form

$$\begin{pmatrix} \Pi_{12} & I \\ \Pi_{22} & 0 \end{pmatrix}$$

in the notation used in Lemma 5 of Section 4.6. This is indeed of full column rank if Π_{22} is of full column rank, which is guaranteed by the rank condition. Hence if the rank condition holds for *each* equation, the conclusion of the lemma follows.

The only importance of the requirement that Σ be nonsingular is that it rules out the inclusion of identities (whose disturbances are identically zero). However, identities typically contain no unknown parameters anyway; they can be dropped from the system, and 3SLS can be applied to the remainder of the system.

7. THEOREM: If all equations are identified by exclusion restrictions and if Σ is nonsingular, the 3SLS estimator is consistent. Furthermore, the asymptotic distribution of $\sqrt{T}(\hat{\alpha} - \alpha)$ is multivariate normal with mean zero and covariance matrix

$$\text{plim}\left[\frac{1}{T} Z_*'(\Sigma^{-1} \otimes X(X'X)^{-1}X')Z_*\right]^{-1}$$

Proof: We begin by showing that T times the matrix H defined in Lemma 5 has a finite probability limit. Note that

$$TH = \left[\frac{1}{T} Z_*'(S^{-1} \otimes X(X'X)^{-1}X')Z_*\right]^{-1} \frac{Z_*'(I \otimes X)}{T} \left[S^{-1} \otimes \left(\frac{X'X}{T}\right)^{-1}\right]$$

The term which is inverted has been shown in Lemma 6 to have a finite and nonsingular probability limit, and it should also be clear that the probability limits of the other two terms are finite. Therefore

$$\bar{H} = \text{plim } TH$$
$$= \text{plim}\left[\frac{1}{T} Z_*'(\Sigma^{-1} \otimes X(X'X)^{-1}X')Z_*\right]^{-1} \frac{Z_*'(I \otimes X)}{T} \left[\Sigma^{-1} \otimes \left(\frac{X'X}{T}\right)^{-1}\right]$$

Now, from Lemma 5 we have that

$$\hat{\alpha} - \alpha = H(I \otimes X')\varepsilon_* = TH \frac{(I \otimes X')\varepsilon_*}{T}$$

But

$$\text{plim } \frac{(I \otimes X')\varepsilon_*}{T} = 0$$

according to Corollary 3 of Section 4.3. Hence $\text{plim}(\hat{\alpha} - \alpha) = 0$, which establishes consistency.

To establish the asymptotic distribution of the estimator, note that

$$\sqrt{T}(\hat{\alpha} - \alpha) = TH \frac{(I \otimes X')\varepsilon_*}{\sqrt{T}}$$

But according to Theorem 2 of Section 4.3,

$$\frac{(I \otimes X')\varepsilon_*}{\sqrt{T}} = \text{vec } \frac{X'\varepsilon}{\sqrt{T}}$$

converges in distribution to $N[0, \Sigma \otimes \text{plim } X'X/T]$. Therefore $\sqrt{T}(\hat{\alpha} - \alpha)$ converges in distribution to a multivariate normal distribution with zero mean and covariance matrix

$$\bar{H}\left(\Sigma \otimes \text{plim } \frac{X'X}{T}\right)\bar{H}'$$
$$= \text{plim}\left\{\frac{1}{T} Z_*'[\Sigma^{-1} \otimes X(X'X)^{-1}X']Z_*\right\}^{-1} .$$

$$\cdot \frac{Z_*'(I \otimes X)}{T}\left[\Sigma^{-1} \otimes \left(\frac{X'X}{T}\right)^{-1}\right]\left(\Sigma \otimes \frac{X'X}{T}\right)\left[\Sigma^{-1} \otimes \left(\frac{X'X}{T}\right)^{-1}\right] \cdot \frac{(I \otimes X')Z_*}{T}$$

$$\cdot \left\{ \frac{1}{T} Z_*'[\Sigma^{-1} \otimes X(X'X)^{-1}X']Z_* \right\}^{-1}$$

$$= \text{plim} \left\{ \frac{1}{T} Z_*'[\Sigma^{-1} \otimes X(X'X)^{-1}X']Z_* \right\}^{-1}$$

8. THEOREM: The 3SLS estimator is efficient relative to the 2SLS estimator.

Proof: What must be shown is that the difference between the asymptotic covariance matrix of the 2SLS estimator and that of the 3SLS estimator is positive semi-definite. The asymptotic covariance matrix of the 2SLS estimator (of all the equations) has already been derived in Proposition 9 of Section 4.6. However, we will rederive it in a somewhat different form in order to make the present comparison easier.

As noted earlier, the 2SLS estimator is

$$\hat{\alpha}_{2SLS} = \{Z_*'[I \otimes X(X'X)^{-1}X']Z_*\}^{-1}Z_*'[I \otimes X(X'X)^{-1}X']y_*$$

Hence

$$\sqrt{T}(\hat{\alpha}_{2SLS} - \alpha) = \{\frac{1}{T} Z_*'[I \otimes X(X'X)^{-1}X']Z_*\}^{-1} \frac{Z_*'(I \otimes X)}{T} \times \left[I \otimes \left(\frac{X'X}{T}\right)^{-1}\right] \frac{(I \otimes X')\varepsilon_*}{\sqrt{T}}$$

Since the asymptotic distribution of

$$\frac{(I \otimes X')\varepsilon_*}{\sqrt{T}}$$

is

$$N(0, \Sigma \otimes \text{plim} \frac{X'X}{T})$$

the asymptotic distribution of $\sqrt{T}(\hat{\alpha}_{2SLS} - \alpha)$ is normal with zero mean and with covariance matrix

$$\operatorname{plim}\left\{\frac{1}{T} Z_*'[I \otimes X(X'X)^{-1}X']Z_*\right\}^{-1} \frac{1}{T} Z_*'[\Sigma \otimes X(X'X)^{-1}X']Z_* \times \left\{\frac{1}{T} Z_*'[I \otimes X(X'X)^{-1}X']Z_*\right\}^{-1}$$

It is therefore sufficient, to show that 3SLS is efficient relative to 2SLS, to show that the above matrix equals the asymptotic covariance matrix of 3SLS *plus* a positive semidefinite matrix.

Let us define the matrix A by

$$\begin{aligned}&\{Z_*'[I \otimes X(X'X)^{-1}X']Z_*\}^{-1}Z_*'(I \otimes X)[I \otimes (X'X)^{-1}]\\ &\quad = \{Z_*'[\Sigma^{-1} \otimes X(X'X)^{-1}X']Z_*\}^{-1}Z_*'(I \otimes X)[\Sigma^{-1} \otimes (X'X)^{-1}] + A\end{aligned}$$

Now, if we consider the expression derived above for the asymptotic covariance matrix of the 2SLS estimator, it is,

$$\begin{aligned}&\operatorname{plim} T\{Z_*'[I \otimes X(X'X)^{-1}X']Z_*\}^{-1}Z_*'[\Sigma \otimes X(X'X)^{-1}X']Z_* \times \{Z_*[I \otimes X(X'X)^{-1}X']Z_*\}^{-1}\\
&= \operatorname{plim} T\left(\{Z_*'[I \otimes X(X'X)^{-1}X']Z_*\}^{-1}Z_*'(I \otimes X)[I \otimes (X'X)^{-1}]\right)(\Sigma \otimes X'X)\\
&\qquad \cdot \left([I \otimes (X'X)^{-1}](I \otimes X')Z_*\{Z_*'[I \otimes X(X'X)^{-1}X']Z_*\}^{-1}\right)\\
&= \operatorname{plim} T\left(\{Z_*'[\Sigma^{-1} \otimes X(X'X)^{-1}X']Z_*\}^{-1}Z_*'(I \otimes X) + A(\Sigma \otimes X'X)\right)\\
&\qquad \cdot \left(A' + [\Sigma^{-1} \otimes (X'X)^{-1}](I \otimes X')Z_*\{Z_*'[\Sigma^{-1} \otimes X(X'X)^{-1}X']Z_*\}^{-1}\right)\\
&= \operatorname{plim} T\{Z_*'[\Sigma^{-1} \otimes X(X'X)^{-1}X']Z_*\}^{-1} + \operatorname{plim} TA(\Sigma \otimes X'X)A'\\
&\qquad + \operatorname{plim} TA(I \otimes X')Z_*\{Z_*'[\Sigma^{-1} \otimes X(X'X)^{-1}X']Z_*\}^{-1}\\
&\qquad + \operatorname{plim} T\{Z_*'[\Sigma^{-1} \otimes X(X'X)^{-1}X']Z_*\}^{-1}Z_*'(I \otimes X)A'\end{aligned}$$

It is easily verified that the last two terms above are zero, in light of the definition of A. Also, the first term is the asymptotic covariance matrix of the 3SLS estimator. Therefore, the difference between the asymptotic covariance matrices of the 2SLS and 3SLS estimators is

$$\operatorname{plim}(TA)\left(\Sigma \otimes \frac{X'X}{T}\right)(TA)'$$

Since Σ and $Q = \text{plim}(X'X/T)$ are both positive definite,

$$\text{plim}\left(\Sigma \otimes \frac{X'X}{T}\right) = \Sigma \otimes Q$$

is positive definite. Hence, the above matrix is positive semi-definite, as was to be proved.

9. COMMENT: If Σ is a diagonal matrix (that is, if there is no correlation across equations), then

$$\{Z_*'[\Sigma^{-1} \otimes X(X'X)^{-1}X']Z_*\}^{-1}Z_*'(I \otimes X)[\Sigma^{-1} \otimes (X'X)^{-1}]$$
$$= \{Z_*'[I \otimes X(X'X)^{-1}X']Z_*\}^{-1}Z_*'(I \otimes X)[I \otimes (X'X)^{-1}]$$

so that $A = 0$. Hence in this case 2SLS and 3SLS are equally efficient. Note, however, that the 2SLS and 3SLS estimation procedures will *not* give identical estimates; even though Σ is diagonal, S will not be. Of course, if it were known a priori that Σ were diagonal, so that only the diagonal elements of S were estimated (the off-diagonal elements being set to zero), then it would be the case that 2SLS and 3SLS would give identical estimates.

3SLS and Identification

We have already seen that each equation must be identified by exclusion restrictions (i.e., each equation must satisfy the usual rank condition) if the 3SLS estimator is to be consistent. We now show that the estimator does not even exist if any equation fails the order condition for identifiability.

10. PROPOSITION: If any equation fails the order condition for identifiability, the 3SLS estimator does not exist.

Proof: For the 3SLS estimator to exist, the matrix

$$Z_*'[S^{-1} \otimes X(X'X)^{-1}X']Z_* = \hat{Z}_*'(S^{-1} \otimes I)\hat{Z}_*$$

must be nonsingular. This requires that the matrix $\hat{Z}_*$ be of full column rank, which requires that each diagonal block

$$\hat{Z}_i = (\hat{Y}_i, X_i) \qquad (i = 1, 2, \cdots, G)$$

be of full column rank. Finally, if the i-th equation does not satisfy the order condition, $\hat{Z}_i$ will not be of full column rank (this is proved in Proposition 10 of Section 4.6), and hence the 3SLS estimator will not exist.

One may note, however, that the underidentified equations can simply be dropped, and 3SLS applied to the remaining equations.

11. THEOREM: If all equations are exactly identified, then 3SLS and 2SLS are (numerically) *identical*.

Proof: By definition, the 3SLS estimator is

$$\hat{\alpha} = \left\{ \begin{pmatrix} Z_1'X & & 0 \\ & \ddots & \\ 0 & & Z_G'X \end{pmatrix} [S^{-1} \otimes (X'X)^{-1}] \begin{pmatrix} X'Z_1 & & 0 \\ & \ddots & \\ 0 & & X'Z_G \end{pmatrix} \right\}^{-1}$$

$$\cdot \begin{pmatrix} Z_1'X & & 0 \\ & \ddots & \\ 0 & & Z_G'X \end{pmatrix} [S^{-1} \otimes (X'X)^{-1}] \begin{pmatrix} X'y_1 \\ \vdots \\ X'y_G \end{pmatrix}$$

Now, if all equations are exactly identified, each diagonal block $X'Z_i$ (or $Z_i'X$) is square and is nonsingular with probability one. We therefore have

$$\hat{\alpha} = \begin{pmatrix} (X'Z_1)^{-1} & & 0 \\ & \ddots & \\ 0 & & (X'Z_G)^{-1} \end{pmatrix} [S \otimes X'X] \begin{pmatrix} (Z_1'X)^{-1} & & 0 \\ & \ddots & \\ 0 & & (Z_G'X)^{-1} \end{pmatrix}$$

$$\cdot \begin{pmatrix} Z_1'X & & 0 \\ & \ddots & \\ 0 & & Z_G'X \end{pmatrix} [S^{-1} \otimes (X'X)^{-1}] \begin{pmatrix} X'y_1 \\ \vdots \\ X'y_G \end{pmatrix}$$

$$= \begin{pmatrix} (X'Z_1)^{-1}X'y_1 \\ \vdots \\ (X'Z_G)^{-1}X'y_G \end{pmatrix}$$

Hence the 3SLS estimator of the parameters of the i-th equation is, in this case, $(X'Z_i)^{-1}X'y_i$.

On the other hand, the 2SLS estimator of the parameters of the i-th equation is

$$[Z_i'X(X'X)^{-1}X'Z_i]^{-1}Z_i'X(X'X)^{-1}X'y_i$$

which in this case equals

$$(X'Z_i)^{-1}X'X(Z_i'X)^{-1}Z_i'X(X'X)^{-1}X'y_i = (X'Z_i)^{-1}X'y_i$$

Thus the 2SLS and 3SLS estimators are identical in this case.

12. COMMENT: Since in the just-identified case, 2SLS is identical to LIML and ILS, 3SLS is also identical to LIML and ILS when all equations are exactly identified.

13. THEOREM: Suppose that the first p equations are overidentified, and that the last G - p are exactly identified. Consider the transformed set of equations

$$(I \otimes X')y_* = (I \otimes X')Z_*\alpha + (I \otimes X')\varepsilon_*$$

and partition it into two sets of equations:

$$\begin{pmatrix} X'y_1 \\ \vdots \\ X'y_p \end{pmatrix} = \begin{pmatrix} X'Z_1 & & 0 \\ & \ddots & \\ 0 & & X'Z_p \end{pmatrix}\begin{pmatrix} \beta_1 \\ \vdots \\ \beta_p \end{pmatrix} + \begin{pmatrix} X'\varepsilon_{\cdot 1} \\ \vdots \\ X'\varepsilon_{\cdot p} \end{pmatrix}$$

$$\begin{pmatrix} X'y_{p+1} \\ \vdots \\ X'y_G \end{pmatrix} = \begin{pmatrix} X'Z_{p+1} & & 0 \\ & \ddots & \\ 0 & & X'Z_G \end{pmatrix}\begin{pmatrix} \beta_{p+1} \\ \vdots \\ \beta_G \end{pmatrix} + \begin{pmatrix} X'\varepsilon_{\cdot p+1} \\ \vdots \\ X'\varepsilon_{\cdot G} \end{pmatrix}$$

In compact notation, write the two sets of equations as

$$r_A = R_A\alpha_A + \eta_A$$

$$r_B = R_B\alpha_B + \eta_B$$

Let S be as previously defined, and partition it conformably as

$$S = \begin{pmatrix} S_{AA} & S_{AB} \\ S_{BA} & S_{BB} \end{pmatrix}$$

and similarly,

$$S^{-1} = \begin{pmatrix} S^{AA} & S^{AB} \\ S^{BA} & S^{BB} \end{pmatrix}$$

Then the 3SLS estimator of α_A is

$$(\hat{\alpha}_A)_{3SLS} = \{R_A'[S_{AA}^{-1} \otimes (X'X)^{-1}]R_A\}^{-1}R_A'[S_{AA}^{-1} \otimes (X'X)^{-1}]r_A$$

while the 3SLS estimator of α_B is

$$(\hat{\alpha}_B)_{3SLS} = (\hat{\alpha}_B)_{2SLS} + R_B^{-1}[(S^{BB})^{-1}S^{BA} \otimes I][r_A - R_A(\hat{\alpha}_A)_{3SLS}]$$

Proof: 3SLS applied to the entire set of equations is the solution to

$$\begin{pmatrix} R_A' & 0 \\ 0 & R_B' \end{pmatrix} \left[\begin{pmatrix} S^{AA} & S^{AB} \\ S^{BA} & S^{BB} \end{pmatrix} \otimes (X'X)^{-1} \right] \begin{pmatrix} R_A & 0 \\ 0 & R_B \end{pmatrix} \begin{pmatrix} \hat{\alpha}_A \\ \hat{\alpha}_B \end{pmatrix}$$

$$= \begin{pmatrix} R_A' & 0 \\ 0 & R_B' \end{pmatrix} \left[\begin{pmatrix} S^{AA} & S^{AB} \\ S^{BA} & S^{BB} \end{pmatrix} \otimes (X'X)^{-1} \right] \begin{pmatrix} r_A \\ r_B \end{pmatrix}$$

This yields two sets of equations:

$$R_A'[S^{AA} \otimes (X'X)^{-1}]R_A\hat{\alpha}_A + R_A'[S^{AB} \otimes (X'X)^{-1}]R_B\hat{\alpha}_B$$
$$= R_A'[S^{AA} \otimes (X'X)^{-1}]r_A + R_A'[S^{AB} \otimes (X'X)^{-1}]r_B \qquad (1)$$

$$R_B'[S^{BA} \otimes (X'X)^{-1}]R_A\hat{\alpha}_A + R_B'[S^{BB} \otimes (X'X)^{-1}]R_B\hat{\alpha}_B$$
$$= R_B'[S^{BA} \otimes (X'X)^{-1}]r_A + R_B'[S^{BB} \otimes (X'X)^{-1}]r_B \qquad (2)$$

Noting that R_B is square and nonsingular with probability one (since the last G - p equations are exactly identified), Eq. (2) can be premultiplied by $(R_B')^{-1}$ to get

$$[S^{BA} \otimes (X'X)^{-1}]R_A\hat{\alpha}_A + [S^{BB} \otimes (X'X)^{-1}]R_B\hat{\alpha}_B$$
$$= [S^{BA} \otimes (X'X)^{-1}]r_A + [S^{BB} \otimes (X'X)^{-1}]r_B$$

This can be solved for $\hat{\alpha}_B$, giving

$$\hat{\alpha}_B = R_B^{-1}r_B + R_B^{-1}[(S^{BB})^{-1}S^{BA} \otimes I](r_A - R_A\hat{\alpha}_A) \tag{3}$$

This proves the second part of the theorem, since $R_B^{-1}r_B$ is indeed the 2SLS estimator of $\hat{\alpha}_B$, given the exact identification of the last G - p equations.

To derive the expression for $\hat{\alpha}_A$, we now substitute (3) into (1):

$$\begin{aligned} R_A'[S^{AA} \otimes (X'X)^{-1}]R_A\hat{\alpha}_A &+ R_A'[S^{AB} \otimes (X'X)^{-1}] \\ &\times \{R_B^{-1}r_B + R_B^{-1}[(S^{BB})^{-1}S^{BA} \otimes I](r_A - R_A\hat{\alpha}_A)\} \\ &= R_A'[S^{AA} \otimes (X'X)^{-1}]r_A + R_A'[S^{AB} \otimes (X'X)^{-1}]r_B \end{aligned}$$

or

$$\begin{aligned} R_A'\{[S^{AA} - S^{AB}(S^{BB})^{-1}S^{BA}] \otimes (X'X)^{-1}\}R_A\hat{\alpha}_A \\ = R_A'\{[S^{AA} - S^{AB}(S^{BB})^{-1}S^{BA}] \otimes (X'X)^{-1}\}r_A \end{aligned} \tag{4}$$

We now wish to show that

$$S^{AA} - S^{AB}(S^{BB})^{-1}S^{BA} = S_{AA}^{-1}$$

To do so we use the partitioned inverse rule:

$$\begin{pmatrix} E & F \\ G & H \end{pmatrix}^{-1} = \begin{pmatrix} D^{-1} & -D^{-1}FH^{-1} \\ -H^{-1}GD^{-1} & (I + H^{-1}GD^{-1}F)H^{-1} \end{pmatrix}$$

where $D = E - FH^{-1}G$. We let

$$\begin{pmatrix} E & F \\ G & H \end{pmatrix} = S^{-1} = \begin{pmatrix} S^{AA} & S^{AB} \\ S^{BA} & S^{BB} \end{pmatrix}$$

Then

$$S_{AA} = [S^{AA} - S^{AB}(S^{BB})^{-1}S^{BA}]^{-1}$$

or

$$S_{AA}^{-1} = S^{AA} - S^{AB}(S^{BB})^{-1}S^{BA}$$

Substituting this in (4) above gives

$$R_A'[S_{AA}^{-1} \otimes (X'X)^{-1}]R_A\hat{\alpha}_A = R_A'[S_{AA}^{-1} \otimes (X'X)^{-1}]r_A$$

and solving this for $\hat{\alpha}_A$ proves the rest of the theorem.

14. COMMENT: In light of the definitions of R_A and r_A above, the expression for the 3SLS estimator of the overidentified equations is

$$\hat{\alpha}_A = \{Z_A'[S_{AA}^{-1} \otimes X(X'X)^{-1}X']Z_A\}^{-1}Z_A'[S_{AA}^{-1} \otimes X(X'X)^{-1}X']y_A$$

where of course Z_A is the submatrix of Z_*, and y_A the subvector of y_*, corresponding to the first p equations. It is then sometimes claimed that this just amounts to applying 3SLS to the overidentified equations only, ignoring the just-identified equations. This is not quite true, since some of the predetermined variables in X may appear only in the just-identified equations. But it is true that the just-identified equations affect the 3SLS estimates of the overidentified equations only by supplying predetermined variables to X.

Also, note that the 3SLS estimate of the parameters of the just-identified equations ($\hat{\alpha}_B$) differs from the 2SLS estimate only by a linear combination of the 3SLS residuals from the overidentified equations.

5.3 FULL INFORMATION MAXIMUM LIKELIHOOD

We again consider the model $Y\Gamma + X\Delta + \varepsilon = 0$, or $Z\beta + \varepsilon = 0$.

1. DEFINITION: The *full information maximum likelihood* (FIML) estimators of the parameters in the model $Y\Gamma + X\Delta + \varepsilon = 0$ are the values of Γ, Δ, and Σ which maximize the joint (log) likelihood function

$$L = -\frac{1}{2}GT\log(2\pi) - \frac{T}{2}\log|\Sigma| + T\log||\Gamma|| - \frac{1}{2}\,\text{trace}\,\Sigma^{-1}(Y\Gamma + X\Delta)'(Y\Gamma + X\Delta)$$

subject to all a priori restrictions.

The computation of the FIML estimates is a nonlinear problem involving a considerable computational burden. However, in the common case in which there are no a priori restrictions on Σ, the following result is of some help.

2. LEMMA: Suppose that there are no a priori restrictions on Σ. Let $\hat{\Sigma}$ be the FIML estimator of Σ, and $\hat{\beta}$ the FIML estimator of β. Then

$$\hat{\Sigma} = \frac{1}{T}\,\hat{\beta}'Z'Z\hat{\beta}$$

Proof: We will maximize the likelihood function

$$L = C_1 + \frac{T}{2}\log|\Sigma^{-1}| + T\log||\Gamma|| - \frac{1}{2}\,\text{trace}\,\Sigma^{-1}\beta'Z'Z\beta$$

with respect to Σ^{-1}. Note that

$$\frac{\partial \log|\Sigma^{-1}|}{\partial \sigma^{ij}} = \sigma_{ji} = \sigma_{ij}$$

and that

$$\frac{\partial\,\text{trace}\,\Sigma^{-1}(\beta'Z'Z\beta)}{\partial\sigma^{ij}} = \frac{\partial \sum_{k=1}^{G}\sum_{m=1}^{G}\sigma^{km}(\beta'Z'Z\beta)_{mk}}{\partial\sigma^{ij}}$$

$$= (\beta'Z'Z\beta)_{ji}$$

$$= (\beta'Z'Z\beta)_{ij}$$

Hence, performing the maximization yields

$$\frac{\partial L}{\partial\sigma^{ij}} = \frac{T}{2}\hat{\sigma}_{ij} - \frac{1}{2}(\hat{\beta}'Z'Z\hat{\beta})_{ij} = 0 \qquad (i,\ j = 1,\ 2,\ \cdots,\ G)$$

In matrix form this is

$$\frac{T}{2}\hat{\Sigma} = \frac{1}{2}\,\hat{\beta}'Z'Z\hat{\beta}$$

or

$$\hat{\Sigma} = \frac{1}{T}\,\hat{\beta}'Z'Z\hat{\beta}$$

3. THEOREM: When Σ is unrestricted (i.e., there are no a priori restrictions on Σ), the FIML estimate of

$$\beta = \begin{pmatrix} \Gamma \\ \Delta \end{pmatrix}$$

is the value which maximizes the concentrated likelihood function

$$L^* = C_2 + T \log||\Gamma|| - \frac{T}{2} \log|\beta' \frac{Z'Z}{T} \beta|$$

subject to the a priori restrictions on β.

Proof: The concentrated likélihood function L^* is obtained by substituting the maximized value of Σ,

$$\Sigma = \frac{1}{T} \beta'Z'Z\beta$$

into the original (log) likelihood function given above. This gives

$$\begin{aligned} L^* &= C_1 - \frac{T}{2} \log|\beta' \frac{Z'Z}{T} \beta| + T \log||\Gamma|| \\ &\quad - \frac{1}{2} \operatorname{trace}[\beta' \frac{Z'Z}{T} \beta]^{-1} \beta'Z'Z\beta \\ &= C_2 - \frac{T}{2} \log|\beta' \frac{Z'Z}{T} \beta| + T \log||\Gamma|| \end{aligned}$$

since

$$\operatorname{trace}\left(\beta' \frac{Z'Z}{T} \beta\right)^{-1} \beta'Z'Z\beta = \operatorname{trace} TI_G = GT \qquad \text{(a constant)}$$

It is perhaps worth noting in passing that, when Σ is unrestricted, the FIML estimator can also be derived by minimizing the generalized residual variance in the reduced-form equations. The reduced form is of course

$$Y = X\Pi + V$$

Now, the generalized residual variance $|V'V|$ is minimized by OLS. Suppose, however, that we minimize it subject to the constraint that $\Pi\Gamma = -\Delta$. Since

$$Z\beta = \varepsilon$$

and

$$V = -\varepsilon\Gamma^{-1},$$

we have

$$|V'V| = |(\Gamma^{-1})'\beta'Z'Z\beta\Gamma^{-1}| = ||\Gamma||^{-2}|\beta'Z'Z\beta|$$

Since minimizing $|V'V|$ is equivalent to maximizing $-(T/2)\log|V'V|$, we may as well maximize

$$-\frac{T}{2}\log|V'V| = T\log||\Gamma|| - \frac{T}{2}\log|\beta'Z'Z\beta|$$

But maximizing this is equivalent to maximizing L^*, as FIML does.

4. COMMENT: Note that FIML involves maximizing the likelihood function *subject to all a priori restrictions*. The most common case is that of only exclusion restrictions on β.

The exclusion restrictions on β can be imposed in the expression for the likelihood function itself. Write each equation as

$$y_i = Z_i\beta_i + \varepsilon_{\cdot i}$$

where

$$Z_i = (Y_i, X_i) \qquad \beta_i = \begin{pmatrix} \gamma_i \\ \delta_i \end{pmatrix}$$

Then the system of equations can be written as

$$y_* = Z_*\alpha + \varepsilon_*$$

just as in our discussion of 3SLS.

Now, since

$$\text{trace } \Sigma^{-1}\beta'Z'Z\beta = (y_* - Z_*\alpha)'(\Sigma^{-1} \otimes I)(y_* - Z_*\alpha)$$

we may write the likelihood function as

$$L = C_1 - \frac{T}{2}\log|\Sigma| + T\log||\Gamma||$$
$$- \frac{1}{2}(y_* - Z_*\alpha)'(\Sigma^{-1} \otimes I)(y_* - Z_*\alpha)$$

FIML consists of maximizing this with respect to Σ and α, noting that some elements of Γ are elements of α, and the rest are either minus one or zero.

When there are no restrictions on Σ, the estimate of Σ (as given by Lemma 2) can be expressed as the matrix S defined by

$$S_{ij} = \frac{1}{T}(y_i - Z_i\beta_i)'(y_j - Z_j\beta_j)$$

The *concentrated* likelihood function (as given by Theorem 3) then becomes

$$L^* = C_2 + T\log||\Gamma|| - \frac{T}{2}\log|S|$$

This notation will prove useful in deriving the asymptotic distribution of the FIML estimator.

The Asymptotic Distribution of the FIML Estimator

In order to write out the asymptotic distribution of the FIML estimators, we need a little further notation. The parameters to be estimated are Γ, Δ, and Σ. If the exclusion restrictions are imposed on Γ and Δ, the unknown elements to be estimated can be expressed as the vector α, as above. We can also express Σ as a vector

$$\text{vec}\,\Sigma = \begin{pmatrix} \Sigma_{\cdot 1} \\ \vdots \\ \Sigma_{\cdot G} \end{pmatrix}$$

Finally, since $\sigma_{ij} = \sigma_{ji}$ for all i and j, we can eliminate some (redundant) elements of vec Σ. Hence let η be the vector of distinct elements of vec Σ, that is

$$\eta = \begin{pmatrix} \sigma_{11} \\ \vdots \\ \sigma_{1G} \\ \sigma_{22} \\ \vdots \\ \sigma_{2G} \\ \sigma_{33} \\ \vdots \\ \sigma_{3G} \\ \vdots \\ \sigma_{GG} \end{pmatrix}$$

In this notation, the unknown parameters are just α and η.

5. PROPOSITION: Let $\hat{\alpha}$ and $\hat{\eta}$ be the FIML estimators of α and η. Then if all equations are identified, $\hat{\alpha}$ and $\hat{\eta}$ are consistent estimators of α and η, and the asymptotic distribution of

$$\sqrt{T}\begin{pmatrix} \hat{\alpha} - \alpha \\ \hat{\eta} - \eta \end{pmatrix}$$

is normal with zero mean and covariance matrix

$$\lim_{T\to\infty}\left\{-\frac{1}{T}E\begin{pmatrix} L_{\alpha\alpha'} & L_{\alpha\eta'} \\ L_{\eta\alpha'} & L_{\eta\eta'} \end{pmatrix}\right\}^{-1} = \left\{-\operatorname{plim}\frac{1}{T}\begin{pmatrix} L_{\alpha\alpha'} & L_{\alpha\eta'} \\ L_{\eta\alpha'} & L_{\eta\eta'} \end{pmatrix}\right\}^{-1}$$

where L is the log likelihood function and the subscripts denote partial differentiation.

Proof: The consistency of the estimates, and the fact that their asymptotic covariance matrix equals the Cramer-Rao lower bound, as given above, are general properties of maximum likelihood estimates. See Theorem 19 of Section A.2, and Comment 20 of Section A.2, which point out that the normal distribution satisfies the required regularity conditions.

The stipulation that all equations be identified is necessary for regularity condition (v) (Comment 20, Section A.2) to be satisfied, however. If any equation is not identified, there exist other sets of parameters which lead to the same value of the likelihood function as do the true parameters, for *any* values of the predetermined variables.

6. COROLLARY: The FIML estimator is asymptotically efficient.

When Σ is not restricted, the asymptotic covariance matrix of the unknown coefficients α can be expressed in terms of the information matrix derived from the concentrated log likelihood function L^*.

7. PROPOSITION: If Σ is not restricted, the asymptotic distribution of $\sqrt{T}(\hat{\alpha} - \alpha)$ is normal with zero mean and with covariance matrix

$$\left[- \operatorname{plim} \frac{1}{T} L^*_{\alpha\alpha'} \right]^{-1}$$

Proof: According to Proposition 5, the covariance matrix of the asymptotic distribution of $\sqrt{T}(\hat{\alpha} - \alpha)$ is the upper left-hand submatrix of

$$\left\{ - \operatorname{plim} \frac{1}{T} \begin{pmatrix} L_{\alpha\alpha'} & L_{\alpha\eta'} \\ L_{\eta\alpha'} & L_{\eta\eta'} \end{pmatrix} \right\}^{-1}$$

Now, we use the partitioned inverse rule, which says that the upper left-hand submatrix of

$$\begin{pmatrix} E & F \\ G & H \end{pmatrix}^{-1}$$

is $(E - FH^{-1}G)^{-1}$; see the proof of Theorem 13 of Section 5.2. Using this result, it follows that the covariance matrix in question is

$$\{- \operatorname{plim} \frac{1}{T} [L_{\alpha\alpha'} - L_{\alpha\eta'}(L_{\eta\eta'})^{-1} L_{\eta\alpha'}]\}^{-1}$$

We wish to show that this equals

$$\left(- \operatorname{plim} \frac{1}{T} L^*_{\alpha\alpha'}\right)^{-1}$$

To do so it is sufficient to show that

$$L^*_{\alpha\alpha'} = L_{\alpha\alpha'} - L_{\alpha\eta} L^{-1}_{\eta\eta'} L_{\eta\alpha'}$$

To show this, recall that to form L^*, we first maximized L with respect to Σ (here, η). That is, we had

$$L_\eta = \frac{\partial L(Y;\alpha,\eta)}{\partial \eta} = 0$$

This was solved for Σ (here, η) to get the maximized value

$$\eta = g(Y;\alpha)$$

Substituting this back into L gives the concentrated likelihood function

$$L^*(Y;\alpha) = L(Y;\alpha,g(Y;\alpha))$$

We can then note that

$$\frac{\partial L^*}{\partial \alpha_i} = \frac{\partial L}{\partial \alpha_i} + \sum_k \frac{\partial L}{\partial \eta_k} \frac{\partial \eta_k}{\partial \alpha_i}$$

$$\frac{\partial^2 L^*}{\partial \alpha_i\, \partial \alpha_j} = \frac{\partial^2 L}{\partial \alpha_i\, \partial \alpha_j} + \sum_k \frac{\partial^2 L}{\partial \alpha_i\, \partial \eta_k} \frac{\partial \eta_k}{\partial \alpha_j}$$

$$+ \sum_k \frac{\partial \eta_k}{\partial \alpha_i} \left[\frac{\partial^2 L}{\partial \eta_k\, \partial \alpha_j} + \sum_m \frac{\partial^2 L}{\partial \eta_k\, \partial \eta_m} \frac{\partial \eta_m}{\partial \alpha_j} \right]$$

$$+ \sum_k \frac{\partial L}{\partial \eta_k} \frac{\partial^2 \eta_k}{\partial \alpha_i\, \partial \alpha_j}$$

The last term here equals zero since $L_\eta = 0$. We can write the rest in matrix form, which gives

$$L^*_{\alpha\alpha'} = L_{\alpha\alpha'} + L_{\alpha\eta'} g_\alpha + g'_\alpha L_{\eta\alpha'} + g'_\alpha L_{\eta\eta'} g_\alpha + 0$$

where the i,j-th element of g_α is

$$\frac{\partial \eta_i}{\partial \alpha_j}$$

Finally, note that $L_\eta = 0$ for any α; differentiating this with respect to α (as above) gives

$$L_{\eta\alpha'} + L_{\eta\eta'}g_\alpha = 0$$

or

$$g_\alpha = -L_{\eta\eta'}^{-1}L_{\eta\alpha'}$$

Substituting this in the expression for $L^*_{\alpha\alpha'}$ above gives

$$L^*_{\alpha\alpha'} = L_{\alpha\alpha'} - L_{\alpha\eta'}L_{\eta\eta'}^{-1}L_{\eta\alpha'}$$

the desired result.

8. THEOREM: When Σ is not restricted, the asymptotic distributions of the 3SLS and FIML estimators of α are identical.

Proof: The asymptotic distribution of the 3SLS estimator was given in Theorem 7 of Section 5.2; the asymptotic distribution of the FIML estimator is given in Proposition 7 above. We therefore need to evaluate

$$\left(-\operatorname{plim}\frac{1}{T}L^*_{\alpha\alpha}\right)^{-1}$$

and to show that it equals the corresponding covariance matrix for the 3SLS estimator.

From Comment 4, we have

$$L^* = C_2 + T\log||\Gamma|| - \frac{T}{2}\log|S|$$

where

$$S_{ij} = \frac{1}{T}(y_i - Z_i\beta_i)'(y_j - Z_j\beta_j)$$

In taking derivatives, we will consider L^* in two terms. First, let

$$P = \frac{\partial^2 T\log||\Gamma||}{\partial\alpha\,\partial\alpha'}$$

Now, we have

$$\frac{\partial T \log||\Gamma||}{\partial \alpha_i} = \begin{cases} 0 & \text{if } \alpha_i \text{ is part of } \Delta \\ T\gamma^{kh} & \text{if } \alpha_i = \gamma_{hk} \end{cases}$$

and

$$p_{ij} = \frac{\partial^2 T \log||\Gamma||}{\partial \alpha_i \, \partial \alpha_j} = \begin{cases} 0 & \text{if either } \alpha_i \text{ or } \alpha_j \text{ is part of } \Delta \\ -T\gamma^{km}\gamma^{nh} & \text{if } \alpha_i = \gamma_{hk} \text{ and } \alpha_j = \gamma_{mn} \end{cases}$$

where the facts used are that

$$\frac{\partial \log|A|}{\partial a_{ij}} = a^{ji}$$

and

$$\frac{\partial a^{ij}}{\partial a_{hk}} = -a^{ih}a^{kj}$$

The second term to be considered is

$$Q = \frac{\partial^2 [-(T/2)\log|S|]}{\partial \alpha \, \partial \alpha'}$$

To calculate this, it is convenient to derive submatrices corresponding to subvectors of α. So consider

$$q_u = \frac{\partial [-(T/2)\log|S|]}{\partial \beta_u}$$

$$= -\frac{T}{2} \sum_i \sum_j \frac{\partial \log|S|}{\partial S_{ij}} \frac{\partial S_{ij}}{\partial \beta_u}.$$

But

$$\frac{\partial \log|S|}{\partial S_{ij}} = S^{ji} = S^{ij}$$

and

$$\frac{\partial S_{ij}}{\partial \beta_u} = \frac{1}{T}\begin{cases} 0 & \text{if } u \neq i \text{ or } j \\ -Z_i'\varepsilon_{\cdot j} & \text{if } u = i, \text{ not } j \\ -Z_j'\varepsilon_{\cdot i} & \text{if } u = j, \text{ not } i \\ -Z_i'\varepsilon_{\cdot j} - Z_j'\varepsilon_{\cdot i} & \text{if } u = i = j \end{cases}$$

Hence

$$q_u = -\frac{T}{2}\left(-\frac{2}{T}\sum_i S^{ui}Z_u'\varepsilon_{\cdot i}\right) = \sum_i S^{ui}Z_u'\varepsilon_{\cdot i}$$

Now, Q consists of submatrices

$$Q_{uv} = \frac{\partial^2 Q}{\partial \beta_u\, \partial \beta_v'} = \frac{\partial q_u}{\partial \beta_v{}'}$$

$$= \frac{\partial \sum_i S^{ui}Z_u'\varepsilon_{\cdot i}}{\partial \beta_v'}$$

$$= \sum_i S^{ui}\frac{\partial Z_u'\varepsilon_{\cdot i}}{\partial \beta_v'} + \sum_i Z_u'\varepsilon_{\cdot i}\sum_j\sum_k \frac{\partial S^{ui}}{\partial S_{jk}}\frac{\partial S_{jk}}{\partial \beta_v'}$$

But

$$\frac{\partial Z_u'\varepsilon_{\cdot i}}{\partial \beta_v'} = \frac{\partial Z_u'(y_i - Z_i\beta_i)}{\partial \beta_v'} = \begin{cases} -Z_u'Z_i & \text{if } v = i \\ 0 & \text{otherwise} \end{cases}$$

so that

$$\sum_i S^{ui}\frac{\partial Z_u'\varepsilon_{\cdot i}}{\partial \beta_v'} = -S^{uv}Z_u'Z_v$$

Also, since

$$\frac{\partial S^{ui}}{\partial S_{jk}} = -S^{uj}S^{ki}$$

we have

$$Q_{uv} = -S^{uv}Z_u'Z_v - \sum_i Z_u'\varepsilon_{\cdot i}\sum_j\sum_k S^{uj}S^{ki}\frac{\partial S_{jk}}{\partial \beta_v'}$$

But

$$\sum_j \sum_k S^{uj} S^{ki} \frac{\partial S_{jk}}{\partial \beta_v'} = -\frac{1}{T} \sum_j S^{uj} S^{vi} \varepsilon_{\cdot j}' Z_v - \frac{1}{T} \sum_k S^{uv} S^{ki} \varepsilon_{\cdot k}' Z_v$$

where Eq. (1) is used, where the first term corresponds to $v = k$, and the second corresponds to $v = j$. Thus

$$\sum_j \sum_k S^{uj} S^{ki} \frac{\partial S_{jk}}{\partial \beta_v'} = -\frac{1}{T} \sum_j (S^{uj} S^{vi} + S^{uv} S^{ji}) \varepsilon_{\cdot j}' Z_v$$

Substituting this above gives

$$Q_{uv} = -S^{uv} Z_u' Z_v + \frac{1}{T} \sum_i \sum_j (S^{uj} S^{vi} + S^{uv} S^{ij}) Z_u' \varepsilon_{\cdot i} \varepsilon_{\cdot j}' Z_v$$

Finally, to write this in somewhat more compact notation, define the $G \times G$ matrix F_{uv} by

$$(F_{uv})_{ij} = S^{uj} S^{vi} + S^{uv} S^{ij}$$

and let

$$\varepsilon = (\varepsilon_{\cdot 1}, \cdots, \varepsilon_{\cdot G})$$

as usual. Then

$$Q_{uv} = -S^{uv} Z_u' Z_v + \frac{1}{T} Z_u' \varepsilon F_{uv} \varepsilon' Z_v$$

We have now derived P and Q; the point is of course that we wish to evaluate

$$-\operatorname{plim} \frac{1}{T} L^*_{\alpha\alpha'} = -\operatorname{plim} \frac{1}{T} (P + Q)$$

Hence we still need to evaluate

$$\operatorname{plim} \frac{1}{T} P \quad \text{and} \quad \operatorname{plim} \frac{1}{T} Q$$

Since P is nonstochastic, the first of these is no problem. We therefore turn to the task of evaluating

$$\text{plim}\ \frac{1}{T} Q$$

We will again consider Q_{uv}, the u,v-th submatrix of Q. We partition this as

$$\text{plim}\ \frac{1}{T} Q_{uv} = \text{plim}\left(\frac{-S^{uv}}{T} \begin{bmatrix} Y_u'Y_v & Y_u'X_v \\ X_u'Y_v & X_u'X_v \end{bmatrix}\right)$$

$$+ \frac{1}{T^2}\begin{pmatrix} Y_u'\varepsilon F_{uv}\varepsilon' Y_v & Y_u'\varepsilon F_{uv}\varepsilon' X_v \\ X_u'\varepsilon F_{uv}\varepsilon' Y_v & X_u'\varepsilon F_{uv}\varepsilon' X_v \end{pmatrix}$$

Now,

$$\text{plim}\ \frac{Y_u'\varepsilon}{T} F_{uv} \frac{\varepsilon' X_v}{T} = 0$$

$$\text{plim}\ \frac{X_u'\varepsilon}{T} F_{uv} \frac{\varepsilon' Y_v}{T} = 0$$

$$\text{plim}\ \frac{X_u'\varepsilon}{T} F_{uv} \frac{\varepsilon' X_v}{T} = 0$$

$$\text{plim}\ S^{uv} = \sigma^{uv}$$

so we have

$$\text{plim}\ \frac{1}{T} Q_{uv} = \text{plim}\begin{pmatrix} -\sigma^{uv} \frac{Y_u'Y_v}{T} + \frac{Y_u'\varepsilon}{T} F_{uv} \frac{\varepsilon' Y_v}{T} & -\sigma^{uv} \frac{Y_u'X_v}{T} \\ -\sigma^{uv} \frac{X_u'Y_v}{T} & -\sigma^{uv} \frac{X_u'X_v}{T} \end{pmatrix}$$

In this expression, the only real problem is the term

$$-\sigma^{uv} \frac{Y_u'Y_v}{T} + \frac{Y_u'\varepsilon}{T} F_{uv} \frac{\varepsilon' Y_v}{T} \equiv \frac{1}{T} B$$

A typical (scalar) element of

$$\text{plim}\ \frac{1}{T} B$$

say the 1,2-nd element, is

$$\text{plim}\ \frac{1}{T} B_{1,2} = -\sigma^{uv}\ \text{plim}\ \frac{1}{T} (\bar{y}_1 + V_1)(\bar{y}_2 + V_2)$$

$$+ \left(\text{plim}\ \frac{1}{T} y_1'\varepsilon\right)\left(\text{plim}\ F_{uv}\right)\left(\text{plim}\ \frac{1}{T} \varepsilon' y_2\right)$$

where $\bar{y}_1$ is $E(y_1)$, as given by the reduced form, and V_1 is the reduced form disturbance.

Now,

$$\text{plim}\ \frac{1}{T} V_1'V_2$$

is the 1,2-nd element of the reduced-form covariance matrix

$$(\Gamma^{-1})'\Sigma\Gamma^{-1}$$

that is,

$$\text{plim}\ \frac{1}{T} V_1'V_2 = \sum_i \sum_j \gamma^{i1}\sigma_{ij}\gamma^{j2}$$

Hence we have

$$\text{plim}\ \frac{1}{T} B_{12} = -\sigma^{uv}\ \text{plim}\ \frac{1}{T} \bar{y}_1'\bar{y}_2 - \sigma^{uv} \sum_i \sum_j \gamma^{i1}\sigma_{ij}\gamma^{j2}$$

$$+ \left(\text{plim}\ \frac{1}{T} y_1'\varepsilon\right)\left(\text{plim}\ F_{uv}\right)\left(\text{plim}\ \frac{1}{T} \varepsilon' y_2\right)$$

The last term of this expression is harder. Again, we can note that

$$\text{plim}\ \frac{1}{T} y_1'\varepsilon$$

is a subvector of $(-\Gamma^{-1})'\Sigma$, so that

$$\left(\text{plim}\ \frac{1}{T} y_1'\varepsilon\right)\left(\text{plim}\ F_{uv}\right)\left(\text{plim}\ \frac{1}{T} \varepsilon' y_2\right)$$

$$= \sum_i \sum_j [(\Gamma^{-1})'\Sigma]_{1i}(\sigma^{uj}\sigma^{vi} + \sigma^{uv}\sigma^{ij})[(\Gamma^{-1})'\Sigma]_{j2}$$

$$= \sum_i \sum_j (\sigma^{uj}\sigma^{vi} + \sigma^{uv}\sigma^{ij})\left(\sum_k \gamma^{k1}\sigma_{ki}\right)\left(\sum_m \gamma^{m2}\sigma_{mj}\right)$$

$$= \sum_i \sum_j (\sigma^{uj}\sigma^{vi} + \sigma^{uv}\sigma^{ij}) \sum_k \sum_m \gamma^{k1}\gamma^{m2}\sigma_{ki}\sigma_{mj}$$

This is, in turn, two terms. The second is

$$\sigma^{uv} \sum_i \sum_j \sum_k \sum_m \sigma^{ij}\gamma^{k1}\gamma^{m2}\sigma_{ki}\sigma_{mj} = \sigma^{uv} \sum_k \sum_m \gamma^{k1}\gamma^{m2} \sum_i \sum_j \sigma_{ki}\sigma^{ij}\sigma_{mj}$$
$$= \sigma^{uv} \sum_k \sum_m \gamma^{k1}\gamma^{m2}\sigma_{km}$$

since

$$\sum_i \sum_j \sigma_{ki}\sigma^{ij}\sigma_{mj}$$

is the k,m-th element of $\Sigma\Sigma^{-1}\Sigma = \Sigma$.

On the other hand, the first term above is

$$\sum_k \sum_m \gamma^{k1}\gamma^{m2}\Big(\sum_i \sigma^{vi}\sigma_{ik}\Big)\Big(\sum_j \sigma^{uj}\sigma_{jm}\Big) = \gamma^{v1}\gamma^{u2}$$

since

$$\sum_i \sigma^{vi}\sigma_{ik} = \begin{cases} 1 & \text{if } k = v \\ 0 & \text{otherwise} \end{cases}$$

$$\sum_i \sigma^{uj}\sigma_{jm} = \begin{cases} 1 & \text{if } m = u \\ 0 & \text{otherwise} \end{cases}$$

Hence we have

$$\Big(\text{plim}\ \frac{1}{T}\ y_1'\varepsilon\Big)\Big(\text{plim}\ F_{uv}\Big)\Big(\text{plim}\ \frac{1}{T}\ \varepsilon'y_2\Big) = \sigma^{uv} \sum_k \sum_m \gamma^{k1}\gamma^{m2}\sigma_{km} + \gamma^{v1}\gamma^{u2}$$

Finally, this implies that

$$\text{plim}\ \frac{1}{T}\ B_{12} = -\sigma^{uv}\ \text{plim}\ \frac{1}{T}\ \bar{y}_1'\bar{y}_2 + \gamma^{v1}\gamma^{u2}$$

noting that two terms have cancelled each other to get this. And this can be written as

$$\text{plim}\ \frac{1}{T}\ B_{12} = -\sigma^{uv}\ \text{plim}\ \frac{1}{T}\ y_1'X(X'X)^{-1}X'y_2 + \gamma^{v1}\gamma^{u2}$$

Now, let us partition (1/T)P as we did Q. If we consider P_{uv}, the submatrix corresponding to the u and v-th equations, note that it is zero except for its upper left-hand block (call it A), of the same size as B. A contains terms of the form $-\gamma^{km}\gamma^{nh}$ corresponding to $\alpha_i = \gamma_{hk}$ and $\alpha_j = \gamma_{mn}$. In particular, the 1,2-nd element of A corresponds to k = u, n = v, h = 1, m = 2; that is, to the first γ coefficient in the u-th equation and the second γ coefficient in the v-th equation. Hence

$$A_{12} = -\gamma^{u2}\gamma^{v1}$$

But this precisely cancels the last term of $\text{plim}(1/T)B_{12}$.

As a result, we can see that the u,v-th block of

$$-\ \text{plim}\ \frac{1}{T}\ (P + Q) = -\ \text{plim}\ \frac{1}{T}\ L^*_{\alpha\alpha'}$$

is just

$$\sigma^{uv}\ \text{plim}\ \frac{1}{T}\begin{pmatrix} Y_u'X(X'X)^{-1}X'Y_v & Y_u'X_v \\ X_u'Y_v & X_u'X_v \end{pmatrix}$$

But this is also the u,v-th block of the matrix which, when inverted, is the covariance matrix of the asymptotic distribution of $\sqrt{T}(\hat{\alpha} - \alpha)_{3SLS}$; see Theorem 7 of Section 5.2. Hence the FIML and 3SLS estimators of α do indeed have the same asymptotic distribution in this case.

9. COROLLARY: When there are no a priori restrictions on Σ, the 3SLS estimator of α is asymptotically efficient.

10. PROPOSITION: When there are a priori restrictions on Σ, the 3SLS estimator of α is less efficient than the FIML estimator.

Proof: We will consider the case in which Σ is known a priori; the principle is the same even when one's knowledge of Σ is not total.

According to Proposition 5, when Σ is *not* known, the covariance matrix of the asymptotic distribution of $\sqrt{T}(\hat{\alpha} - \alpha)_{FIML}$ is the upper submatrix of

$$\left(-\ \text{plim}\ \frac{1}{T}\begin{bmatrix} L_{\alpha\alpha'} & L_{\alpha\eta'} \\ L_{\eta\alpha'} & L_{\eta\eta'} \end{bmatrix}\right)^{-1}$$

Equivalently (see the proof of Proposition 7), it is

$$\left[- \operatorname{plim} \frac{1}{T} (L_{\alpha\alpha'} - L_{\alpha\eta'} L_{\eta\eta'}^{-1} L_{\eta\alpha'})\right]^{-1}$$

As shown in Theorem 8, this is also the covariance matrix of the asymptotic distribution of $\sqrt{T}(\hat{\alpha} - \alpha)_{3SLS}$.

On the other hand, if Σ is known, the relevant covariance matrix becomes simply

$$(- \operatorname{plim} \frac{1}{T} L_{\alpha\alpha'})^{-1}$$

Clearly, these two expressions are equal if and only if

$$\operatorname{plim} \frac{1}{T} L_{\alpha\eta'} = 0$$

To check whether or not this is the case, we again write the log likelihood function as

$$L = C_1 + \frac{T}{2} \log|\Sigma^{-1}| + T \log||\Gamma||$$
$$- \frac{1}{2} (y_* - Z_*\alpha)'(\Sigma^{-1} \otimes I)(y_* - Z_*\alpha)$$

Hence

$$\frac{\partial L}{\partial \alpha} = T \frac{\partial \log||\Gamma||}{\partial \alpha} + Z_*'(\Sigma^{-1} \otimes I)(y_* - Z_*\alpha)$$

We will now differentiate this with respect to an element of Σ^{-1} [see Note 1]. The first (cross-partial) term will clearly equal zero. Also, note that

$$Z_*'(\Sigma^{-1} \otimes I)(y_* - Z_*\alpha) = \begin{pmatrix} \sum_j \sigma^{1j} Z_1'\varepsilon_{\cdot j} \\ \vdots \\ \sum_j \sigma^{Gj} Z_G'\varepsilon_{\cdot j} \end{pmatrix}$$

Hence, for example,

$$\frac{\partial^2 L}{\partial\alpha\, \partial\sigma^{11}} = \begin{pmatrix} Z_1'\varepsilon_{\cdot 1} \\ 0 \\ \vdots \\ 0 \end{pmatrix}$$

Furthermore,

$$\text{plim}\, \frac{1}{T}\, Z_1'\varepsilon_{\cdot 1} \neq 0$$

Indeed, this is just the "simultaneous equations problem."

Hence, we have

$$\text{plim}\, \frac{1}{T}\, L_{\alpha\eta'} \neq 0$$

The estimates of Σ and of β are not asymptotically independent, and knowledge of Σ improves the asymptotic efficiency of estimation of β.

11. COMMENT: Note that the difference between the asymptotic distributions of 3SLS and FIML when Σ is known (or restricted) depends on the presence of endogenous variables among the regressors. As we saw in Chapter 2, in the context of seemingly unrelated regressions (with no such endogenous variables), knowledge of Σ did not help the asymptotic efficiency of the coefficient estimates.

The inefficiency of 3SLS when Σ is known is apparently due to the fact that Σ is involved not only in the correlation of disturbances across equations but also in the correlation of the endogenous regressors with the disturbances. 3SLS merely "purges" the endogenous regressors of their correlation with the disturbances; apparently this is inefficient when there is a priori knowledge about this correlation.

Linearized FIML

The main difficulty with the FIML estimation technique is that it involves a large amount of computation--it is a nonlinear technique. We now show that FIML can be "linearized" without loss of asymptotic efficiency.

12. DEFINITION: Let L be the log likelihood function of the sample, θ the vector of parameters to be estimated, and $\hat{\theta}^o$ an initial consistent estimate of $\hat{\theta}$, such that $\sqrt{T}(\hat{\theta}^o - \theta)$ has a well-defined asymptotic distribution. Then the *linearized maximum likelihood estimator* of θ is

$$\hat{\theta}^L = \hat{\theta}^o - \left(\frac{\partial^2 L}{\partial\theta\, \partial\theta'} \right)^{-1}_{\theta=\hat{\theta}^o} \left(\frac{\partial L}{\partial\theta} \right)_{\theta=\hat{\theta}^o}$$

13. PROPOSITION: The linearized maximum likelihood estimator of θ has the same asymptotic distribution as the maximum likelihood estimator.

Proof: As a matter of definition, we have

$$\hat{\theta}^L_i = \hat{\theta}^o_i - \sum_k G_{ik} \left(\frac{\partial L}{\partial\theta_k} \right)_{\theta=\hat{\theta}^o}$$

where G_{ik} is the i,k-th element of

$$\left(\frac{\partial^2 L}{\partial\theta\, \partial\theta'} \right)^{-1}_{\theta=\hat{\theta}^o}$$

Also note that, if $\hat{\theta}^M$ is the maximum likelihood estimator of θ, then

$$\left(\frac{\partial L}{\partial\theta_k} \right)_{\theta=\hat{\theta}^M} = 0$$

Expanding this by Taylor series around $\hat{\theta}^o$, we get

$$\left(\frac{\partial L}{\partial\theta_k} \right)_{\theta=\hat{\theta}^o} + \sum_m (\hat{\theta}_m{}^M - \hat{\theta}^o_m) \left(\frac{\partial^2 L}{\partial\theta_k\, \partial\theta_m} \right)_{\theta=\hat{\theta}^o}$$

$$+ \frac{1}{2} \sum_m \sum_n (\hat{\theta}_m{}^M - \hat{\theta}^o_m)(\hat{\theta}_n{}^M - \hat{\theta}^o_n) \left(\frac{\partial^3 L}{\partial\theta_k\, \partial\theta_m\, \partial\theta_n} \right)_{\theta=\theta^*} = 0$$

where θ^* is between $\hat{\theta}^M$ and $\hat{\theta}^o$. Solving this for

$$\left(\frac{\partial L}{\partial\theta_k} \right)_{\theta=\hat{\theta}^o}$$

and inserting it in the first equation above gives

$$\hat{\theta}_i^{\,L} = \hat{\theta}_i^o + \sum_k G_{ik} \left\{ \sum_m (\hat{\theta}_m^{\,M} - \hat{\theta}_m^o) \left(\frac{\partial^2 L}{\partial\theta_m \, \partial\theta_k} \right)_{\theta=\hat{\theta}^o} \right.$$

$$\left. + \frac{1}{2} \sum_m \sum_n (\hat{\theta}_m^{\,M} - \hat{\theta}_m^o)(\hat{\theta}_n^{\,M} - \hat{\theta}_n^o) \left(\frac{\partial^3 L}{\partial\theta_k \, \partial\theta_m \, \partial\theta_n} \right)_{\theta=\theta^*} \right\}$$

Now, the first term in brackets amounts to

$$\sum_m (\hat{\theta}_m^{\,M} - \hat{\theta}_n^o) \sum_k G_{ik} \left(\frac{\partial^2 L}{\partial\theta_k \, \partial\theta_m} \right)_{\theta=\hat{\theta}^o}$$

But

$$\sum_k G_{ik} \left(\frac{\partial^2 L}{\partial\theta_k \, \partial\theta_m} \right)_{\theta=\hat{\theta}^o} = \begin{cases} 1 & \text{if } i = m \\ 0 & \text{otherwise} \end{cases}$$

so that the above term becomes simply $\hat{\theta}_i^{\,M} - \hat{\theta}_i^o$. Hence we have

$$\hat{\theta}_i^{\,L} = \hat{\theta}_i^{\,M} + \frac{1}{2} \sum_k G_{ik} \sum_m \sum_n (\hat{\theta}_m^{\,M} - \hat{\theta}_m^o)(\hat{\theta}_n^{\,M} - \hat{\theta}_n^o) \left(\frac{\partial^3 L}{\partial\theta_k \, \partial\theta_m \, \partial\theta_n} \right)_{\theta=\hat{\theta}_*}$$

This can be written as

$$\hat{\theta}_i^{\,L} = \hat{\theta}_i^{\,M} + (\hat{\theta}^M - \hat{\theta}^o)' A_i (\hat{\theta}^M - \hat{\theta}^o)$$

where the m,n-th element of the matrix A_i is

$$(A_i)_{m,n} = \frac{1}{2} \sum_k G_{ik} \left(\frac{\partial^3 L}{\partial\theta_k \, \partial\theta_m \, \partial\theta_n} \right)_{\theta=\hat{\theta}^*}$$

Hence we have

$$\sqrt{T}(\hat{\theta}^L - \theta) = \sqrt{T}(\hat{\theta}^M - \theta) + \Delta$$

where the i-th element of Δ is

$$\Delta_i = \sqrt{T}(\hat{\theta}^M - \hat{\theta}^o)' A_i (\hat{\theta}^M - \hat{\theta}^o)$$

In order to prove the proposition, therefore, all that needs to be shown is that plim $\Delta = 0$. But this is clear because

$$\sqrt{T}(\hat{\theta}^M - \hat{\theta}^o) = \sqrt{T}(\hat{\theta}^M - \theta) - \sqrt{T}(\hat{\theta}^o - \theta)$$

has a well-defined asymptotic distribution, and plim A_i is well-defined [see Note 2], while

$$\text{plim}(\hat{\theta}^M - \hat{\theta}^o) = 0$$

in light of the consistency of both $\hat{\theta}^M$ and $\hat{\theta}^o$.

14. COMMENT: The above result is a general one; its application is not limited to the present context of linearizing the FIML technique. Also, it should be clear that the technique can be carried out iteratively; given the estimate of $\hat{\theta}^L$, it can be used as the "initial consistent estimate" $\hat{\theta}^o$, *another* estimate $\hat{\theta}^L$ calculated, and so forth. Such iterations do not gain any asymptotic efficiency, however.

5.4 THE DERIVED REDUCED FORM

We return to the complete model written as $Y\Gamma + X\Delta + \varepsilon = 0$. The reduced form is $Y = X\Pi + V$, where $\Pi = -\Delta\Gamma^{-1}$ and $V = -\varepsilon\Gamma^{-1}$. The rows of V are iid as $N(0,\Omega)$ where $\Omega = (\Gamma^{-1})'\Sigma\Gamma^{-1}$, and where Σ is the covariance matrix of the rows of ε.

We have already seen (Theorem 6 of Section 4.3) that the ordinary least squares estimator of Π,

$$\hat{\Pi} = (X'X)^{-1}X'Y$$

is consistent, and that

$$\sqrt{T}\ \text{vec}(\hat{\Pi} - \Pi) \rightarrow N(0,\Omega \otimes \underset{\sim}{Q}^{-1})$$

where

$$\underset{\sim}{Q} = \text{plim}\ \frac{1}{T}\ X'X$$

We now consider a different method of estimation. If we have consistent estimates of the structural parameters, say $\tilde{\Gamma}$ and $\tilde{\Delta}$, we estimate Π as

$$\tilde{\Pi} = -\tilde{\Delta}\tilde{\Gamma}^{-1}$$

This estimate will be consistent if $\tilde{\Delta}$ and $\tilde{\Gamma}$ are consistent estimates of Δ and Γ.

1. DEFINITION: Let $\tilde{\Delta}$ and $\tilde{\Gamma}$ be consistent estimators of Δ and Γ. Then the *derived reduced-form* estimator of Π is

$$\tilde{\Pi} = -\tilde{\Delta}\tilde{\Gamma}^{-1}$$

2. COMMENT: When some of the structural equations are overidentified, there are less unknown parameters in Δ and Γ than in Π. The identity

$$\Pi = -\Delta\Gamma^{-1}$$

therefore imposes certain restrictions on Π. The potential advantage of the derived reduced form is that these restrictions are taken into account, whereas the ordinary least squares estimate of Π ignores these implicit restrictions. This raises at least the possibility that the derived reduced-form estimator of Π may be more efficient than the ordinary least squares estimator.

3. THEOREM: Suppose that

$$\sqrt{T}\,\mathrm{vec}\begin{pmatrix} \tilde{\Gamma} - \Gamma \\ \tilde{\Delta} - \Delta \end{pmatrix} \to N(0, \Psi)$$

and let $\tilde{\Pi} = -\tilde{\Delta}\tilde{\Gamma}^{-1}$. Then

$$\sqrt{T}\,\mathrm{vec}(\tilde{\Pi} - \Pi) \to N(0, A\Psi A')$$

where

$$A = (\Gamma^{-1})' \otimes (\Pi, I)$$

Proof:

$$\begin{aligned} \tilde{\Pi} - \Pi &= \tilde{\Pi} + \tilde{\Delta}\Gamma^{-1} - \tilde{\Delta}\Gamma^{-1} - \Pi = \tilde{\Pi}\Gamma\Gamma^{-1} - \tilde{\Pi}\tilde{\Gamma}\Gamma^{-1} - \tilde{\Delta}\Gamma^{-1} + \Delta\Gamma^{-1} \\ &= -\tilde{\Pi}(\tilde{\Gamma} - \Gamma)\Gamma^{-1} - (\tilde{\Delta} - \Delta)\Gamma^{-1} \\ &= -(\tilde{\Pi}, I)\begin{pmatrix} \tilde{\Gamma} - \Gamma \\ \tilde{\Delta} - \Delta \end{pmatrix}\Gamma^{-1} \end{aligned}$$

We now use the fact that vec(ABC) = $(C' \otimes A)\text{vec}(B)$ to get

$$\text{vec}(\tilde{\Pi} - \Pi) = -[(\Gamma^{-1})' \otimes (\tilde{\Pi}, I)]\text{vec}\begin{pmatrix} \tilde{\Gamma} - \Gamma \\ \tilde{\Delta} - \Delta \end{pmatrix}$$

This implies

$$\sqrt{T}\ \text{vec}(\tilde{\Pi} - \Pi) = -\tilde{A}\sqrt{T}\ \text{vec}\begin{pmatrix} \tilde{\Gamma} - \Gamma \\ \tilde{\Delta} - \Delta \end{pmatrix}$$

where

$$\tilde{A} = (\Gamma^{-1})' \otimes (\tilde{\Pi}, I)$$

But then if

$$\sqrt{T}\ \text{vec}\begin{pmatrix} \tilde{\Gamma} - \Gamma \\ \tilde{\Delta} - \Delta \end{pmatrix} \to N(0,\ \Psi)$$

it follows that

$$\sqrt{T}\ \text{vec}(\tilde{\Pi} - \Pi) \to N(0,\ A\Psi A)$$

as claimed.

4. COMMENT: In general Ψ will contain a large number of zeroes, corresponding to the elements of Δ and Γ that are known a priori.

5. PROPOSITION: Let γ_i and δ_i be the unknown parameters in the i-th column of Γ and Δ. Let

$$\alpha = \begin{pmatrix} \gamma_1 \\ \delta_1 \\ \gamma_2 \\ \delta_2 \\ \vdots \\ \gamma_G \\ \delta_G \end{pmatrix}$$

(as in the discussion of 3SLS). Suppose that $\tilde{\alpha}$ is an estimate of α such that

$$\sqrt{T}(\tilde{\alpha} - \alpha) \rightarrow N(0, \Psi_*)$$

(Note that Ψ_* corresponds to Ψ with the zeroes deleted.) Define

$$D_* = (\Gamma^{-1})' \otimes I_K$$

and

$$W_* = \begin{bmatrix} W_1 & & & 0 \\ & W_2 & & \\ & & \ddots & \\ 0 & & & W_G \end{bmatrix}$$

where

$$W_i = \text{plim}(X'X)^{-1}X'(Y_i, X_i)$$

Then

$$\sqrt{T}\ \text{vec}(\tilde{\Pi} - \Pi) \rightarrow N(0, D_*W_*\Psi_*W_*'D_*')$$

Proof: From Theorem 3 we know that

$$\sqrt{T}\ \text{vec}(\tilde{\Pi} - \Pi) \rightarrow N(0, A\Psi A')$$

where $A = (\Gamma^{-1})' \otimes (\Pi, I)$. But

$$(\Gamma^{-1})' \otimes (\Pi, I) = [(\Gamma^{-1})' \otimes I_K][I_G \otimes (\Pi, I)]$$
$$= D_*[I_G \otimes (\Pi, I)]$$

Therefore, to show $A\Psi A' = D_*W_*\Psi_*W_*'D_*'$, we need to show that

$$[I_G \otimes (\Pi, I)]\Psi[I_G \otimes (\Pi, I)]' = W_*\Psi_*W_*'$$

To do so, we note that

$$[I_G \otimes (\Pi, I)]\Psi[I_G \otimes (\Pi, I)']$$

is the asymptotic covariance matrix of

$$[I_G \otimes (\Pi, I)]\sqrt{T}\,\mathrm{vec}\begin{pmatrix} \tilde{\Gamma} - \Gamma \\ \tilde{\Delta} - \Delta \end{pmatrix} = \begin{bmatrix} (\Pi, I)\sqrt{T}\begin{pmatrix} \tilde{\Gamma}_{\cdot 1} - \Gamma_{\cdot 1} \\ \tilde{\Delta}_{\cdot 1} - \Delta_{\cdot 1} \end{pmatrix} \\ \vdots \\ (\Pi, I)\sqrt{T}\begin{pmatrix} \tilde{\Gamma}_{\cdot G} - \Gamma_{\cdot G} \\ \tilde{\Delta}_{\cdot G} - \Delta_{\cdot G} \end{pmatrix} \end{bmatrix}$$

For simplicity, consider the first block element of this vector,

$$(\Pi, I)\sqrt{T}\begin{pmatrix} \tilde{\Gamma}_{\cdot 1} - \Gamma_{\cdot 1} \\ \tilde{\Delta}_{\cdot 1} - \Delta_{\cdot 1} \end{pmatrix}$$

In our usual notation,

$$\Gamma_{\cdot 1} = \begin{pmatrix} -1 \\ \gamma_1 \\ 0 \end{pmatrix} \qquad \Delta_{\cdot 1} = \begin{pmatrix} \delta_1 \\ 0 \end{pmatrix}$$

$$\Pi = \begin{pmatrix} \Pi_{11} & \Pi_{12} & \Pi_{13} \\ \Pi_{21} & \Pi_{22} & \Pi_{23} \end{pmatrix}$$

Then

$$(\Pi, I)\sqrt{T}\begin{pmatrix} \tilde{\Gamma}_{\cdot 1} - \Gamma_{\cdot 1} \\ \tilde{\Delta}_{\cdot 1} \quad \Delta_{\cdot 1} \end{pmatrix} = \begin{pmatrix} \Pi_{12} & I \\ \Pi_{22} & 0 \end{pmatrix}\sqrt{T}\begin{pmatrix} \hat{\gamma}_1 - \gamma_1 \\ \hat{\delta}_1 - \delta_1 \end{pmatrix}$$

$$= [\mathrm{plim}(X'X)^{-1}X'(Y_1, X_1)]\sqrt{T}\begin{pmatrix} \tilde{\gamma}_1 - \gamma_1 \\ \tilde{\delta}_1 - \delta_1 \end{pmatrix}$$

$$= W_1\sqrt{T}\begin{pmatrix} \tilde{\gamma}_1 - \gamma_1 \\ \tilde{\delta}_1 - \delta_1 \end{pmatrix}.$$

Proceeding in the same way with the other block elements of the above vector, we arrive at

$$\sqrt{T}\begin{bmatrix} W_1\begin{pmatrix} \tilde{\gamma}_1 - \gamma_1 \\ \tilde{\delta}_1 - \delta_1 \end{pmatrix} \\ \vdots \\ W_G\begin{pmatrix} \tilde{\gamma}_G - \gamma_G \\ \tilde{\delta}_G - \delta_G \end{pmatrix} \end{bmatrix} = W_*\sqrt{T}(\tilde{\alpha} - \alpha)$$

with W_* and α as defined above. Since $\sqrt{T}(\tilde{\alpha} - \alpha) \rightarrow N(0, \Psi_*)$ $W_*\sqrt{T}(\tilde{\alpha} - \alpha) \rightarrow N(0, W_*\Psi_*W_*')$. This proves the proposition.

6. PROPOSITION: The derived reduced form based on 3SLS structural estimates is asymptotically efficient relative to the OLS estimate of the reduced form.

Proof: The asymptotic covariance of the OLS estimate can be written as

$$\text{plim } T[\Omega \otimes (X'X)^{-1}]$$

The asymptotic covariance matrix of the 3SLS derived reduced form is of the form

$$D_*W_*\Psi_*W_*'D_*'$$

where

$$\Psi_* = \text{plim } T\{Z_*'[\Sigma^{-1} \otimes X(X'X)^{-1}X']Z_*\}^{-1}$$

It is therefore sufficient, to prove the proposition, to show that

$$\text{plim } T\left(\Omega \otimes (X'X)^{-1} - D_*W_*\{Z_*'[\Sigma^{-1} \otimes X(X'X)^{-1}X']Z_*\}^{-1}W_*'D_*'\right)$$

is positive semidefinite.

To proceed, note that

$$\begin{aligned} \Omega \otimes (X'X)^{-1} &= [(\Gamma^{-1})'\Sigma\Gamma^{-1}] \otimes (X'X)^{-1} \\ &= [(\Gamma^{-1})' \otimes I][\Sigma \otimes (X'X)^{-1}](\Gamma^{-1} \otimes I) \\ &= D_*[\Sigma \otimes (X'X)^{-1}]D_*' \end{aligned}$$

Therefore the expression which we hope to show is positive semi-definite is

$$\text{plim}\ TD_*\left(\Sigma \otimes (X'X)^{-1} - W_*\{Z_*'[\Sigma^{-1} \otimes X(X'X)^{-1}X']Z_*\}^{-1}W_*'\right)D_*'$$

This will be positive semidefinite if

$$\text{plim}\ T\left(\Sigma \otimes (X'X)^{-1} - W_*\{Z_*'[\Sigma^{-1} \otimes X(X'X)^{-1}X']Z_*\}^{-1}W_*'\right)$$

is positive semidefinite. We can also note that

$$W_* = \text{plim}[I \otimes (X'X)^{-1}X']Z_*$$

With this substitution we get

$$\text{plim}\ T\left(\Sigma \otimes (X'X)^{-1} - [I \otimes (X'X)^{-1}X']Z_*\{Z_*'[\Sigma^{-1} \otimes X(X'X)^{-1}X']Z_*\}^{-1} \times Z_*'[I \otimes X(X'X)^{-1}]\right)$$

We next premultiply by $Z_*'(I \otimes X)$ and postmultiply by its transpose to obtain

$$\text{plim}\ T\Big(\{Z_*'[\Sigma \otimes X(X'X)^{-1}X']Z_*\} - \{Z_*'[I \otimes X(X'X)^{-1}X']Z_*\}\{Z_*'[\Sigma^{-1} \otimes X(X'X)^{-1}X']Z_*\}^{-1} \times \{Z_*'[I \otimes X(X'X)^{-1}X']Z_*\}\Big)$$

Finally, we premultiply and postmultiply by $\{Z_*'[I \otimes X(X'X)^{-1}X']Z_*\}^{-1}$, which gives

$$\text{plim}\ T\Big(\{Z_*'[I \otimes X(X'X)^{-1}X']Z_*\}^{-1}\{Z_*'[\Sigma \otimes X(X'X)^{-1}X']Z_*\} \times \{Z_*'[I \otimes X(X'X)^{-1}X']Z_*\}^{-1} - \{Z_*'[\Sigma^{-1} \otimes X(X'X)^{-1}X']Z_*\}^{-1}\Big)$$

But this is indeed positive semidefinite since it is the difference between the asymptotic covariance matrix of the 2SLS estimator and that of the 3SLS estimator; see the proof of Theorem 8 of Section 5.2.

7. COMMENT: We have shown that the 3SLS derived reduced form is asymptotically efficient relative to the OLS-estimated reduced form. It follows that the FIML derived reduced form is likewise asymptotically efficient relative to OLS. On the other hand, we have *not* shown that derived reduced forms based on inefficient structural

estimates (e.g., 2SLS) are efficient relative to OLS; indeed, such an assertion would be incorrect.

8. COMMENT: It should also be clear that if all equations are just identified, the derived reduced forms (based on 2SLS, 3SLS, etc.) and OLS-estimated reduced form are *identical*. This is so because when all equations are exactly identified, there is a one-to-one correspondence between reduced-form parameters and unknown structural parameters. Indirect least squares and the derivation of the reduced form from structural estimates are then simply inverse operations.

EXERCISES

1. Suppose that the vector α of unknown structural coefficients is known to satisfy the linear restriction

$$R\alpha = r$$

where R and r are a known matrix and vector, respectively. Let $\hat{\alpha}$ be the 3SLS estimate. Consider the estimator

$$\tilde{\alpha} = \hat{\alpha} + CR'(RCR')^{-1}(r - R\hat{\alpha})$$

where

$$C = \{Z_*'[S^{-1} \otimes X(X'X)^{-1}X']Z_*\}^{-1}$$

Prove that $\tilde{\alpha}$ is consistent and is asymptotically efficient relative to $\hat{\alpha}$.

2. Suppose that one has an estimate $\hat{\Pi}$ of the reduced-form parameters such that

$$\sqrt{T}\, \text{vec}(\hat{\Pi} - \Pi) \rightarrow N(0,\theta)$$

(The form of θ will of course depend on the method used to estimate Π.) Let the subscript f denote some future time period (outside the sample used to estimate Π). We are interested in predicting

$$Y_f = (Y_{f1}, \cdots, Y_{fG})$$

given

$$X_f = (X_{f1}, \cdots, X_{fK})$$

We consider the prediction $\hat{Y}_f = X_f\hat{\Pi}$.

a. Show that

$$\hat{Y}_f - Y_f = X_f(\hat{\Pi} - \Pi) - V_f$$

where

$$V_f = (V_{f1}, \cdots, V_{fG})$$

is the reduced-form disturbance in time period f.

b. Show that

$$\sqrt{T}\ \text{vec}[X_f(\hat{\Pi} - \Pi)] \to N(0, A\Theta A')$$

where

$$A = (I_G \otimes X_f)$$

c. Conclude that the asymptotic covariance matrix of

$$\hat{Y}_f' - Y_f'$$

is

$$\frac{1}{T} A\Theta A' + \Omega$$

3. Consider the estimator $\hat{\alpha} = A'(I \otimes X')y_*$, where A is the solution to

$$\begin{pmatrix} S \otimes X'X & (I \otimes X')Z_* \\ Z_*'(I \otimes X) & 0 \end{pmatrix} \begin{pmatrix} A \\ B \end{pmatrix} = \begin{pmatrix} 0 \\ I \end{pmatrix}$$

Show that this estimator is the 3SLS estimator. Give an interpretation to the solution obtained for B.

4. In the context of Exercise 3, reorder the equations so that the overidentified equations come first and the exactly identified equations come last. Correspondingly partition y_*, Z_*, $\hat{\alpha}$, A, B, and S:

$$y_* = \begin{bmatrix} y_{*1} \\ y_{*2} \end{bmatrix}, \quad Z_* = \begin{bmatrix} Z_{*1} & 0 \\ 0 & Z_{*2} \end{bmatrix},$$

$$\hat{\alpha} = \begin{bmatrix} \hat{\alpha}_1 \\ \hat{\alpha}_2 \end{bmatrix}, \quad A = \begin{bmatrix} A_{11} & A_{12} \\ A_{21} & A_{22} \end{bmatrix},$$

$$B = \begin{bmatrix} B_{11} & B_{12} \\ B_{21} & B_{22} \end{bmatrix}, \quad S = \begin{bmatrix} S_{11} & S_{12} \\ S_{21} & S_{22} \end{bmatrix}$$

The equation defining $\hat{\alpha}$ becomes

$$\begin{bmatrix} \hat{\alpha}_1 \\ \hat{\alpha}_2 \end{bmatrix} = \begin{bmatrix} A'_{11} & A'_{21} \\ A'_{12} & A'_{22} \end{bmatrix} \begin{bmatrix} (I \otimes X')y_{*1} \\ (I \otimes X')y_{*2} \end{bmatrix}$$

while the equations defining A and B become

$$\begin{bmatrix} S_{11} \otimes X'X & S_{12} \otimes X'X & (I \otimes X')Z_{*1} & 0 \\ S_{21} \otimes X'X & S_{22} \otimes X'X & 0 & (I \otimes X')Z_{*2} \\ Z'_{*1}(I \otimes X) & 0 & 0 & 0 \\ 0 & Z'_{*2}(I \otimes X) & 0 & 0 \end{bmatrix}$$

$$\times \begin{bmatrix} A_{11} & A_{12} \\ A_{21} & A_{22} \\ B_{11} & B_{12} \\ B_{21} & B_{22} \end{bmatrix} = \begin{bmatrix} 0 & 0 \\ 0 & 0 \\ I & 0 \\ 0 & I \end{bmatrix}$$

a. Show that $Z'_{*2}(I \otimes X)$ is square and nonsingular.

b. Show that this implies $A_{21} = 0$, and therefore

$$\hat{\alpha}_1 = A'_{11}(I \otimes X')y_{*1}$$

c. Show that the solution for A_{11} may as well be obtained, in light of the fact $A_{21} = 0$, from the equations

$$\begin{pmatrix} S_{11} \otimes X'X & (I \otimes X')Z_{*1} \\ Z_{*1}'(I \otimes X) & 0 \end{pmatrix} \begin{pmatrix} A_{11} \\ B_{11} \end{pmatrix} = \begin{pmatrix} 0 \\ I \end{pmatrix}$$

d. Comparing this to the set of equations in Exercise 3, conclude that 3SLS applied to the overidentified equations only gives the same estimates for those equations as does 3SLS applied to the whole system.

NOTES

Section 5.3

1. It is easier to differentiate this expression with respect to Σ^{-1} than with respect to Σ. In the present context it makes no difference which is done since

$$\frac{\partial^2 L}{\partial \alpha_i \, \partial \sigma^{jk}} = 0 \qquad \text{(for all } j \text{ and } k)$$

if and only if

$$\frac{\partial^2 L}{\partial \alpha_i \, \partial \sigma_{jk}} = 0 \qquad \text{(for all } j \text{ and } k)$$

2. Under the regularity conditions as given in Theorem 19 of Section A.2; see condition (ii).

REFERENCES

Court, R.H. (1974), "Three Stage Least Squares and Some Extensions where the Structural Disturbance Covariance Matrix May Be Singular," *Econometrica 42*, 529-546.

Dhrymes, P.J. (1970), *Econometrics: Statistical Foundations and Applications* New York: Harper and Row.

__________ (1973), "Restricted and Unrestricted Reduced Forms: Asymptotic Distribution and Relative Efficiency," *Econometrica 41*, 119-134.

Goldberger, A.S., A.L. Nagar, and H.S. Odeh (1961), "The Covariance Matrices of Reduced Form Coefficients and of Forecasts for a Structural Econometric Model," *Econometrica 29*, 556-573.

Goldberger, A.S. (1964), *Econometric Theory* New York: John Wiley.

Hood, W.C., and T.C. Koopmans (ed.)(1953), *Studies in Econometric Method* New York: John Wiley.

Narayan, R. (1969), "Computation of Zellner-Theil's Three Stage Least Squares Estimates," *Econometrica 37*, 298-306.

Nissen, D. (1968), "A Note on the Variance of a Matrix," *Econometrica 36*, 603-604.

Rothenberg, T.J., and C.T. Leenders (1964), "Efficient Estimation of Simultaneous Equations Systems," *Econometrica 32*, 57-76.

Theil H. (1971), *Principles of Econometrics* New York: John Wiley.

Zellner, A., and H. Theil (1962), "Three-Stage Least Squares: Simultaneous Estimation of Simultaneous Equations," *Econometrica 30*, 54-78.

Appendix

This appendix presents, without proof, certain basic statistical results which are used in the text. Most of them can be found in a book of the order of difficulty of Wilks (1962).

A.1 CONVERGENCE OF RANDOM VARIABLES

1. DEFINITION: Let $x_1, x_2, \cdots$ be a sequence of (vector) random variables. That is, each x_t is a p-dimensional random vector $(x_{t1}, x_{t2}, \cdots, x_{tp})'$. Then the sequence $x_1, x_2, \cdots$ is said to *converge in probability* to the random variable $z = (z_1, z_2, \cdots, z_p)'$ if for any (arbitrary) $\varepsilon > 0$,

$$\lim_{t\to\infty} P[|x_{ti} - z_i| \geq \varepsilon] = 0 \qquad (i = 1, 2, \cdots, p)$$

We then write $\text{plim } x_t = z$.

Of course, the vector z may be a constant vector (a degenerate random vector). Hence convergence to a constant is just a special case of convergence to a random variable.

2. PROPOSITION: Let $x_1, x_2, \cdots$ be a sequence of random variables such that

$$\lim_{t\to\infty} E(x_{ti} - z_i)^2 = 0 \qquad (i = 1, 2, \cdots, p)$$

Then the sequence x_1, x_2, $\cdots$ converges in probability to the random variable z [see Note 1].

3. THEOREM (Slutsky): Suppose plim $x_t = z$, z a constant, and that g is a continuous function on p-dimensional Euclidean space. Then

$$\text{plim } g(x_t) = g(z) \qquad \text{[see Note 2]}$$

4. DEFINITION: Let x_1, x_2, $\cdots$ be a sequence of random variables with distribution functions F_1, F_2, $\cdots$; and let F be a distribution function. Then the sequence x_1, x_2, $\cdots$ is said to *converge in distribution* to the distribution F if

$$\lim_{t\to\infty} F_t = F$$

at every point of continuity of F.

5. COMMENT: This does *not* imply that the moments of the random variables x_1, x_2, $\cdots$ converge to the moments of the random variable z having distribution function F. For example, it need not be true that

$$\lim_{t\to\infty} E(x_t) = E(z)$$

Indeed, E(z) can exist even though $E(x_t)$ does not exist for any t.

6. THEOREM: If the sequence x_1, x_2, $\cdots$ converges in probability to a random variable z, and if F is the distribution function of z, then the sequence x_1, x_2, $\cdots$ converges in distribution to F [see Note 3].

7. DEFINITION: The *characteristic function* of a random variable x is defined by

$$\phi(\xi) = E[\exp(i\xi'x)]$$

where ξ is an arbitrary vector of real constants and $i = \sqrt{-1}$.

8. THEOREM: Let $x_1, x_2, \cdots$ be a sequence of random variables with characteristic functions $\phi_1, \phi_2, \cdots$. Then a necessary and sufficient condition for this sequence to converge in distribution to some distribution is that

$$\lim_{t\to\infty} \phi_t(\xi) = \phi(\xi) \qquad \text{(for all } \xi)$$

where $\phi(\xi)$ is a function which is continuous at $\xi = 0$. Furthermore, ϕ is then the characteristic function of the distribution to which the sequence converges [see Note 4].

As an example of the use of Theorem 8, note that the characteristic function of the multivariate normal distribution $N(\mu,\Sigma)$ is

$$\phi(\xi) = \exp(i\xi'\mu - \tfrac{1}{2}\xi'\Sigma\xi)$$

Now consider the sequence $x_1, x_2, \cdots$, where x_t is distributed as $N(\mu_t,\Sigma_t)$ and hence has characteristic function

$$\phi_t(\xi) = \exp(i\xi'\mu_t - \tfrac{1}{2}\xi'\Sigma_t\xi)$$

Then if $\lim_{t\to\infty} \mu_t = \mu$ and $\lim_{t\to\infty} \Sigma_t = \Sigma$, it is clear that

$$\lim_{t\to\infty} \phi_t(\xi) = \exp(i\xi'\mu - \tfrac{1}{2}\xi'\Sigma\xi) \qquad \text{(for all } \xi)$$

Hence the sequence converges in distribution to $N(\mu,\Sigma)$.

9. THEOREM: Let $A_1, A_2, \cdots$ be a sequence of random $p \times p$ matrices, and $x_1, x_2, \cdots$ be a sequence of p-dimensional vector random variables. Suppose that $\text{plim } A_t = \bar{A}$ exists, and that the sequence x_t converges in distribution to $N(0,\Sigma)$. Then the sequence $A_t x_t$ converges in distribution to $N(0,\bar{A}\Sigma\bar{A}')$.

A.2 ESTIMATION THEORY

1. DEFINITION: An estimator $\hat{\theta}$ of a parameter θ is a function of observable random variables.

2. DEFINITION: Let $\hat{\theta}_t$ be an estimator of θ which is based on t observations on a set of random variables. Then $\hat{\theta}_t$ is

Unbiased if $E(\hat{\theta}_t) = \theta$

Asymptotically unbiased if $\lim_{t\to\infty} E(\hat{\theta}_t) = \theta$

Consistent if $\text{plim}\ \hat{\theta}_t = \theta$

3. COMMENT: Consistency does not imply asymptotic unbiasedness. Indeed, an estimator can be consistent even though $E(\hat{\theta}_t)$ does not even exist for any t. Nor does asymptotic unbiasedness imply consistency. However, if $\hat{\theta}_t$ is asymptotically unbiased and if its covariance matrix vanishes asymptotically, then it is consistent.

Efficiency

4. DEFINITION: Let $\hat{\theta}$ and $\tilde{\theta}$ be unbiased estimators of a parameter θ, and let their covariance matrices be Σ and Ω, respectively. Then we say that $\hat{\theta}$ *is efficient relative to* $\tilde{\theta}$ if $\Omega - \Sigma$ is a positive semi-definite matrix.

Of course, if $\hat{\theta}$ and $\tilde{\theta}$ are estimators of a scalar parameter, then the above definition just requires that $\text{Var}\ \tilde{\theta} \geq \text{Var}\ \hat{\theta}$.

5. DEFINITION: An estimator $\hat{\theta}$ of a parameter θ is a *best linear unbiased estimator* (BLUE) if

(i) It is unbiased

(ii) It is a linear function of the observed random variables on which it is based

(iii) It is efficient relative to any other estimator satisfying (i) and (ii).

6. DEFINITION: An estimator $\hat{\theta}$ of a parameter θ is *efficient* if it is efficient relative to all other unbiased estimators of θ.

Clearly, efficiency implies best linear unbiasedness. However, best linear unbiasedness is often easier to establish than efficiency. One way to establish efficiency is through the following results.

7. DEFINITION: Let $X = \{x_1, x_2, \cdots, x_T\}$ be a random sample from a population with density $f(\cdot,\theta)$. (That is, the x_t are mutually independent and each has density f.) Then the statistic S(X) is

sufficient for θ if the conditional distribution of X, given the value of S(X), is independent of θ.

8. THEOREM (Neyman-Fisher Criterion): Let $X = \{x_1, x_2, \cdots, x_T\}$ be a random sample from a population with density $f(\cdot,\theta)$. Then the statistic S(X) is sufficient for θ if and only if the likelihood function

$$L(X,\theta) = \prod_{t=1}^{T} f(x_t,\theta)$$

can be factored as

$$L(X,\theta) = g(S(X);\theta)h(X)$$

where g depends on X only through the value of S(X) and h does not depend on θ [see Note 1].

9. THEOREM (Blackwell-Rao): Let S be a sufficient statistic for θ, and suppose that $\tilde{\theta}$ is an unbiased estimator of θ. Then the statistic $\hat{\theta}$ defined as a function of S by the conditional expectation

$$\hat{\theta}(S) = E(\tilde{\theta}|S)$$

is an unbiased estimator which is efficient relative to $\tilde{\theta}$ [see Note 2].

If, as is usually the case, there is only one unbiased estimator of θ which is a function of the sufficient statistic S, then that estimator is efficient by Theorem 9. The following results are one way of ensuring that this is the case.

10. DEFINITION: A sufficient statistic is *complete* if no function of it has zero expectation unless it is zero with probability one.

11. THEOREM: If the distribution $f(\cdot,\theta)$ admits of a complete sufficient statistic S for θ, then an unbiased estimator of θ which is a function of S is efficient [see Note 3].

The above results dealing with sufficient statistics provide one way of showing efficiency. Another way is through the Cramer-Rao inequality, given below.

12. DEFINITION: A density $f(\cdot,\theta)$ is *regular with respect to its first derivatives* if

$$E\left[\frac{\partial \log f(\cdot,\theta)}{\partial \theta_i}\right] = 0 \qquad (i = 1, 2, \cdots, p) \quad \text{[see Note 4]}$$

13. THEOREM (Cramer-Rao): Let $X = \{x_1, x_2, \cdots, x_T\}$ be a random sample from a population with density $f(\cdot,\theta)$ which is regular with respect to its first derivatives. Let the likelihood function be

$$L(X,\theta) = \prod_{t=1}^{T} f(x_t,\theta)$$

Define the information matrix

$$I = -E\left[\frac{\partial^2 \log L(X,\theta)}{\partial\theta\, \partial\theta'}\right]$$

that is,

$$I_{ij} = -E\left[\frac{\partial^2 \log L}{\partial\theta_i\, \partial\theta_j}\right]$$

Suppose that $\hat{\theta}$ is any unbiased estimator, with covariance matrix Σ. Then the matrix $\Sigma - I^{-1}$ is positive semidefinite [see Note 5].

It therefore follows immediately that if an estimator has covariance matrix I^{-1}, it is efficient. However, an estimator may be efficient even though its covariance matrix does not attain this lower bound. In this case efficiency must be shown in some other manner, such as by showing the estimator to be a function of a complete sufficient statistic.

Asymptotic Efficiency

14. DEFINITION: Suppose that $\hat{\theta}$ is a (consistent) estimator of θ, and that $\sqrt{T}(\hat{\theta}_T - \theta)$ converges in distribution to $N(0,\Sigma)$. Then $\hat{\theta}$ is said to have the asymptotic distribution $N(\theta,\Sigma/T)$.

In general, if $\sqrt{T}(\hat{\theta}_T - \theta)$ converges to a distribution with zero mean and covariance matrix Σ, the asymptotic covariance matrix of $\hat{\theta}$ is said to be Σ/T.

15. DEFINITION: An estimator $\hat{\theta}$ is *consistent uniformly asymptotically normal* (CUAN) if it is consistent, if $\sqrt{T}(\hat{\theta}_T - \theta)$ converges in distribution to $N(0,\Sigma)$, and if the convergence is uniform over any compact (closed and bounded) subset of the parameter space.

16. DEFINITION: If $\hat{\theta}$ is a CUAN estimator with asymptotic covariance matrix Σ/T and $\tilde{\theta}$ is a CUAN estimator with asymptotic covariance matrix Ω/T, then $\hat{\theta}$ *is asymptotically efficient relative to* $\tilde{\theta}$ if the matrix $\Omega - \Sigma$ is positive semidefinite.

17. DEFINITION: A CUAN estimator is *asymptotically efficient* if it is asymptotically efficient relative to any other CUAN estimator.

18. COROLLARY (to Cramer-Rao Theorem): A sufficient condition for a CUAN estimator to be asymptotically efficient is that its asymptotic covariance matrix be equal to the lower bound

$$\frac{1}{T} \lim_{T\to\infty} \left(\frac{I}{T}\right)^{-1}$$

The following result is useful since it says that, under certain regularity conditions, maximum likelihood estimators are consistent and asymptotically efficient.

19. THEOREM: Let $X = \{x_1, x_2, \cdots, x_T\}$ be a random sample from a population with density $f(\cdot,\theta)$. Let $\hat{\theta}_T$ be the maximum likelihood estimator of θ. Then $\hat{\theta}_T$ is a consistent estimator of θ, and $\sqrt{T}(\hat{\theta}_T - \theta)$ has asymptotic distribution

$$N[0, \lim_{T\to\infty}(I/T)^{-1}]$$

if the following "regularity conditions" are satisfied [see Note 6]:

(i) The range of the random variables x_t is independent of θ.

(ii) $f(\cdot,\theta)$ possesses third-order derivatives with respect to θ, and these are bounded by integrable functions of x.

(iii) The space of admissible values of θ is a closed subset of p-dimensional cartesian space.

(iv) If $\theta_1, \theta_2, \cdots$ is a sequence such that

$$\lim_{i \to \infty} |\theta_i| = \infty$$

then

$$\lim_{i \to \infty} f(x, \theta_i) = 0$$

for all x.

(v) Let $F(\cdot, \theta)$ be the distribution function corresponding to $f(\cdot, \theta)$. Then if θ is the true parameter value, and θ^* is any other admissible parameter value,

$$F(x, \theta) \neq F(x, \theta^*)$$

for at least one x.

(vi) For the true parameter value θ,

$$E|\log f(x, \theta)| < \infty$$

(vii) For any θ and $\rho > 0$, let $g(x, \theta, \rho)$ be the supremum of $f(x, \theta^*)$ with respect to all θ^* such that $|\theta - \theta^*| \leq \rho$. Let

$$g^*(x, \theta, \rho) = \begin{cases} g(x, \theta, \rho) & \text{if } g(x, \theta, \rho) > 1 \\ 1 & \text{otherwise} \end{cases}$$

Furthermore, let $\phi(x, r)$ be the supremum of $f(x, \theta^*)$ with respect to all θ^* such that $|\theta^*| > r$; let

$$\phi^*(x, r) = \begin{cases} \phi(x, r) & \text{if } \phi(x, r) > 1 \\ 1 & \text{otherwise} \end{cases}$$

Then $E(\log g^*)$ and $E(\log \phi^*)$ are finite for sufficiently small ρ and sufficiently large r. Furthermore, $g(x, \theta, \rho)$ is a measurable function of x for any θ and ρ.

20. COMMENT: The normal distribution satisfies the above regularity conditions.

A.3 CENTRAL LIMIT THEOREMS

1. THEOREM (Lindberg-Levy): Let $x_1, x_2, \cdots,$ be a sequence of independent and identically distributed random variables with finite mean μ and finite variance σ^2. Then the random variable

$$Z_T = \frac{1}{\sqrt{T}} \sum_{t=1}^{T} \frac{x_t - \mu}{\sigma}$$

converges in distribution to N(0,1) [see Note 1].

2. THEOREM (Lindberg-Feller): Let $x_1, x_2, \cdots,$ be a sequence of independent random variables. Suppose that for every t, x_t has finite mean μ_t and variance σ_t^2; let F_t be the distribution function of x_t. Define

$$S_T^2 = \sum_{t=1}^{T} \sigma_t^2$$

Then a necessary and sufficient condition for the random variable

$$Z_T = \sum_{t=1}^{T} \frac{x_t - \mu_t}{S_T}$$

to converge in distribution to N(0,1) is that for every $\varepsilon > 0$,

$$\lim_{T\to\infty} \frac{1}{S_T^2} \sum_{t=1}^{T} \int_{|\omega-\mu_t|>\varepsilon S_T} (\omega-\mu_t)^2 dF_t(\omega) = 0 \qquad \text{[see Note 2]}$$

The following theorem is useful in converting problems of convergence of multivariate random variables to problems of univariate convergence.

3. THEOREM: Let $x_1, x_2, \cdots$ be a sequence of p-dimensional vector random variables. Then a necessary and sufficient condition for the sequence x_t to converge to a distribution F is that the sequence of scalar random variables

$$\lambda' x_t$$

converge in distribution for *any* p-vector λ [see Note 3].

4. COROLLARY: Let $x_1, x_2, \cdots$ be a sequence of vector random variables. Then if the sequence $\lambda' x_t$ converges in distribution to $N(0, \lambda'\psi\lambda)$ for every λ, the sequence x_t converges in distribution to $N(0, \psi)$.

The chief drawback to Theorems 1 and 2 is that they deal only with sequences of independent random variables. We will now allow for a certain type of dependence.

5. DEFINITION: Let $x_1, x_2, \cdots$ be a sequence of random variables. This sequence is said to be *m-dependent* if the two subsequences

$$(x_1, x_2, \cdots, x_r) \qquad (x_s, x_{s+1}, \cdots, x_n)$$

are independent whenever $r + m < s$.

6. COMMENT: 0-dependence is equivalent to independence. m-dependence implies that two elements of the sequence are independent if the difference between their indices is greater than m.

7. THEOREM: Let $x_1, x_2, \cdots$, be a sequence of m-dependent random variables with zero mean and with uniformly bounded third absolute moment, that is,

$$E(x_t) = 0 \quad \text{and} \quad E|x_t|^3 < R \qquad (t = 1, 2, \cdots)$$

Define

$$A_t = E(x_{t+m}^2) + 2 \sum_{j=0}^{m-1} E(x_{i+j} x'_{i+m})$$

and suppose that

$$\sigma^2 = \lim_{H \to \infty} \frac{1}{H} \sum_{h=1}^{H} A_{t+h}$$

exists and is independent of t. Then the random variable

$$z_T = \frac{1}{\sqrt{T}} \sum_{t=1}^{T} x_t$$

converges in distribution to $N(0, \sigma^2)$ [see Note 4].

Finally, the following theorem is particularly useful in many econometric applications.

8. THEOREM: Consider the multivariate autoregressive process

$$y_t = Z_t C_0 + \sum_{i=1}^{H} y_{t-i} C_i + \varepsilon_t$$

where y_t is a $1 \times G$ vector of the t-th observation on the G dependent variables, Z_t is a $1 \times N$ vector of the t-th observation on N non-stochastic independent variables, ε_t is a $1 \times G$ disturbance vector, C_0 is an $N \times G$ coefficient matrix, and C_i $(i = 1, 2, \cdots, H)$ are $G \times G$ coefficient matrices. Assume that

(i) The ε_t' are iid as $N(0,\Sigma)$.

(ii) $|Z_{ti}| < M$, where $i = 1, 2, \cdots, N$; $t = 1, 2, \cdots, \infty$; M being a finite constant.

(iii) $\lim_{T\to\infty} \frac{1}{T-\theta} \sum_{t=1}^{T-\theta} Z_t' Z_{t+\theta}$ exists for all θ, and

$\lim_{T\to\infty} \frac{1}{T} \sum_{t=1}^{T} Z_t' Z_t$ is nonsingular.

(iv) All roots (λ) of the equation

$$\left|\lambda^H I_G - \sum_{h=1}^{H} \lambda^{H-h} C_h\right| = 0$$

are less than one in absolute value (modulus).

Finally, let

$$X_t = (Z_t, y_{t-1}, \cdots, y_{t-H})$$

and let X be the matrix of all T observations, X_t; its dimension is $T \times (N + GH)$. Then

$$\underset{\sim}{Q} = \text{plim}\, \frac{X'X}{T}$$

is finite and nonsingular, and

$$\text{vec}\, \frac{X'\varepsilon}{\sqrt{T}}$$

converges in distribution to $N(0, \Sigma \otimes \underset{\sim}{Q})$ [see Note 5].

NOTES

Section A.1

1. Wilks (1962), p. 100.
2. Wilks (1962), p. 103.
3. Wilks (1962), p. 101.
4. Wilks (1962), p. 122.

Section A.2

1. Wilks (1962), p. 355.
2. Wilks (1962), p. 357. Note that $\hat{\theta}$ is indeed an estimator since it is a function of S(X).
3. Rao (1965), p. 261; Rao (1973), p. 321.
4. A sufficient condition for the condition of Definition 12 is that the derivatives $(\partial/\partial\theta_i)f(x,\theta)$ be bounded by integrable functions $h_i(x)$. See Wilks (1962), pp. 345-350.
5. Rao (1965), p. 265; Rao (1973), p. 326.
6. Dhrymes (1970), pp. 120-121. For slightly less stringent conditions, see Wald (1949).

Section A.3

1. Rao (1965), p. 107; Rao (1973), p. 127; Wilks (1962), p. 256.
2. Rao (1965), p. 108; Rao (1973), p. 128.
3. Rao (1965), p. 108; Rao (1973), p. 128.
4. Hoeffding and Robbins (1948).
5. Schönfeld (1971). Incidentally, for any matrix A, vec(A) is the vector formed by "stacking" its columns. That is, if A = $(A_1, A_2, \cdots, A_n)$, then

$$\text{vec}(A) = \begin{pmatrix} A_1 \\ A_2 \\ \vdots \\ A_n \end{pmatrix}$$

REFERENCES

Dhrymes, P.J. (1970), *Econometrics: Statistical Foundations and Applications* New York: Harper and Row.

Hoeffding, W., and H. Robbins (1948), "The Central Limit Theorem for Dependent Variables," *Duke Mathematical Journal 15*, 773-780.

Rao, C.R. (1965), *Linear Statistical Inference and Its Applications* New York: John Wiley.

Rao, C.R. (1973), *Linear Statistical Inference and Its Applications*, 2nd ed. New York: John Wiley.

Schönfeld, P. (1971), "A Useful Central Limit Theorem for m-Dependent Random Variables," *Metrika 17*, 116-128.

Wald, A. (1949), "Note on the Consistency of the Maximum Likelihood Estimate," *Annals of Mathematical Statistics 20*, 595-601.

Wilks, S.S. (1962), *Mathematical Statistics* New York: John Wiley.

Author Index

Subject Index

J,K,L

M

ISBN 0-8247-8735-8
90000>
EAN
9 780824 787356